INDIAN CHAROPHYTA

MJP PUBLISHERS

INDIAN CHAROPHYTA

D. Subramanian

Professor of Botany (Retd.),
Annamalai University,
Annamalai Nagar – 608 002,
Tamil Nadu.

MJP PUBLISHERS

Chennai **Trichy** **Tirunelveli** **New Delhi**

MJP PUBLISHERS

ISBN 978-81-8094-358-4

MJP Publishers

No. 44, Nallathambi Street,
Triplicane, Chennai 600 005

MJP 377

Publisher: **C. Janarthanan**

Project Editor: **C. Ambica**

This Book is Dedicated to
Shri. Kanagambigai Amman and
Lord Uchinatha Swamy of Sivapuri, Chidambaram,
Tamil Nadu.

Preface

Studies on cytotaxonomical and chemotaxonomical aspects and ecological adaptations leading to evolution of varieties and species of Charophytes of India have been given importance in this book.

Some of the forms, varieties and species of Charophyta as reported from various parts of India and studied by the author also been also added in this book. The author has described a large number of forms, varieties and species of Charophyta from all over India, excluding the members of Charophyta already published in his previous book "Monograph on Indian Charophyta."

A special chapter on Taxonomic validity of *Nitella terrestris* M.O.P. Iyengar has been added in this book.

Another chapter on adaptations of Charophytes on the rain water tracks of sand dunes of Tamil Nadu and origin and evolution of Charophytes has been also included in this book.

Acknowledgement

The author expresses his sincere thanks to Dr. K. Arumugam, Former Professor and Head, Department of Botany, DDE, Annamalai University, Tamilnadu for permitting him to go through the departmental herbariums on algae.

The author is also thankful to Dr. K. Rangasamy, Former Director, CAS in Botany, Madras University, India for permitting him to go through important books and journals in the general library of the Department.

The author is thankful to Dr. D. Dhanavel, Dr. P. Ranganathan, Dr. D. Kumarasamy, Dr. K. Saravanakumar, Associate Professors, DDE Wing and Main Departments of Botany, Annamalai University for having helped him in herbarium verifications and microscopic observations of fresh and herbarium specimens of Charophyta.

The author also expresses his heartful thanks to Dr. Venkatesalu, the present Professor and Head, DDE Wing, Botany Department, Annamalai University for permitting him to examine some of the herbarium specimens of *Chara* and *Nitella*.

Finally, the author is thankful to Mr. Sheshadri Ramaswamy, a lover of plants, Bangalore, for having collected and given him some Charophytes from Bangalore and its neighbouring areas.

D. Subramanian

Introduction

The present contribution on Indian Charophyta is the continuation of the author's previous study "Monograph on Indian Charophyta" in which a large number of new species and varieties, both monoecious and dioecious of *Chara*, *Nitella* and *Tolypella* had been described along with those already described by V.S. Sundaralingam in his paper "A Systematic Account of the South Indian *Characeae*" in 1954. Totally 17 species of *Nitella*, 9 species of *Chara*, 3 varieties coming under 2 species of *Tolypella* along with *Lycnothamnus barbatus* and *Nitellopsis obtusa* had been described in "Monograph on Indian Charophyta" in 2002.

Further studies on the species of Charophyta have been undertaken from various parts of India and they have been described in this book.

In my previous book "Monograph on Indian Charophyta" fundamental studies on developmental morphology of vegetative and reproductive parts of Charophytic members have been furnished. Besides, economic importance, fossil studies, various systems of classification of Charophyta and key preparation for the species of *Chara* and *Nitella* included in this book have been briefly provided.

H.S. Pundhir (1998) has provided an interesting resume on Karyology of Indian Charophyta in "Progress in cytogenetics" edited by. P. Kachroo and published by Bisen Singh Mahendra Pal Singh, Dehradun, India. The more important aspects of this resume have also been added in this book.

The author has made comparative studies on free and protein bound amino acids and chlorophyll a, b and total chlorophyll content among the species of *Chara*, *Nitella*, *Nitellopsis* and *Lychnothamnus* to understand the interrelationship among the members of Charophyta. These chemotaxonomical studies have been also included in this book.

Ecological adaptations of Charophytes in sand dunes of Puduchattiram and its neighbourhood at Cuddalore District of Tamilnadu during rainy seasons have been presented as a special lecture in UGC national seminar on present scenario in plant sciences research "held at Botany Department, AnnamalaiUniversity on February 19 and 20, 2005. The observations of this presentation have been also added in this book.

A large number of forms, varieties and species of Charophytes have been described in this book with illustrations of the author from all over India. Therefore, varieties and species of Charophyta which were included in my previous book "Monograph on Indian Charophyta" are not included in this book.

Contents

I

RECENT STUDIES ON INDIAN CHAROPHYTA

The members of Charophyta, which were included in the previous book "Monograph on Indian Charophyta" are listed below:

1. *Nitella acuminata* A. Br.
2. *N. dualis* var. *pulchella. f superba.* RDW.
3. *N. furcata* subsp. *mucronata* (A.Br.) R.D.W.
4. *N. furcata* subsp. *flagellifera f. patula* (I.B.R. & G.O.A.) R.D.W.
5. *N. pseudoflabellata* var. *imperialis. f. wattii* (J.Gr.) R.D.W.
6. *N. pseudoflabellata* var. *elegans.* R.D.W.
7. *N. pseudoflabellata* var. *elegans.f. spinosa.* R.D.W.
8. *N. furcata* subsp. *mucronata. f. oligospira.* (A.Br.) R.D.W.
9. *N. furcata* subsp. *flagelliformis.* (A.Br.) R.D.W. Combo. Nov.
10. *N. tenuissima f. bengalensis.* (Kundu) R.D.W.
11. *N. tenuissima.* (Desv.) R.D.W.
12. *N. grovesii.* Kundu.
13. *N. gracilis.* var. *athanurensis.* D. Subramanian.
14. *N. gracilis* var. *sathamangalensis.* D. Subramanian.
15. *N. myriotrichia* var. *acuminata.* D. Subramanian.
16. *N. myriotrichia* var. *incurvata.* D. Subramanian.
17. *N. hyalina* var. *chelliammai.* D. Subramanian.

5. *Nitella hyalina* var. *hyalina f. indica* (A. Br.) R.D.W. by Pundhir H.S., Aruna Gautam and Vidyavati (1993).
6. *Charafibrosa* var. *fibrosa f. longicorollata* (Kasaki) R.D.W. by Pundhir, H.S. and Vidyavati (1994).
7. *Charavulgaris* var. *imperfectaf. dissoluta* (A. Br. Ex. Leonh) R.D.W. by Pundhir, H.S. and Vidyavati (1994).
8. *Charahornemannii f. longifolia* (Rob.) R.D.W. by Pundhir, H.S. and Aruna Gautam (1994).
9. *Charavulgaris* var. *vulgaris. f. crispa* (wallm) R.D.W. by Pundhir. H.S. and Aruna Gautam (1994).
10. *Charapseudohydropitys* Imah. by Pundhir. H.S., Aruna Gautam and Vidyavati (1994).

In these 10 species, cytological studies have been also made by Pundhir and his co-workers, who had made extensive field trips to various parts of Uttar Pradesh and Jammu regions. These findings are the latest contributions from Northern parts of India.

II

KARYOLOGY OF INDIAN CHAROPHYTA

I. CHROMOSOME NUMBERS

Khan and Sarma (1967a, b, c) made the most significant contribution on karyology of Indian Charophyta. Earlier, Sundaralingam (1946) studied the cytology of *Charazeylanica* perhaps the first attempt from India. Further additions were those by Noor (1968), Sinha and Noor (1967, 1971), Sinha and Verma (1970), Ahmed and Sinha (1973) Mukherjee and Noor (1973), Sinha and Ahmad (1974, 1979), Noor and Mukherjee (1975, 1977), Prasad and Verma (1985), Labh and Verma (1984, 1985a, b, c) and others from Bihar. Sarma and Ramjee (1967) recorded chromosome members in *Nitellatenuissima* from Arsam. Ramjee and Sarma (1971) further studied the cytology of 41 taxa from various parts of India. Chatterjee (1970, 1976a, b, 1979a) Mukhopadhyay and Chatterjee (1980) and Ray and Chatterjee (1986a, b, 1987, 1988a, b, 1989) studied the cytology of Charophyta from West Bengal. Bharati and Chennaveeraiah (1974, 1980) and Chennaveeraiah and Bharati (1974) from Mysore and Khan and Sarma (1980) worked in *Lamprothamniumsuccintum* and *Nitellaacuminata f. belongeri* from Sarnath, Varanasi. Patel and Jawale (1984) and Jawale and Patil (1985) worked in *Charahaitensis* and *Nitellapolycarpa* from Gujarat respectively and from Uttar Pradesh by Ramjee and Bhatnagar (1978a, b), Bhatnagar (1988a, b, 1989a, b), Bhatnagar and Gangawar (1985), Bhatnagar and Sharma (1985), Bhatnagar and Johri (1986, 1991) and Pundhir and Vidyavati (1993, 1994a). Subramanian (1983) and Subramanian and Ganesan (1983), studied the cytology of some taxa from Tamilnadu, Sarona and Bala (1984) in four taxa of *Chara* from Punjab, Bharati (1990) of two polyplid races of *Charafibrosa* complex. Subramaniyam, *et al.*, (1991) recorded aneuploids in two taxa of *Nitella* from Karnataka and Kerala. Pundhir and Gautam (1993) recorded new chromosome number in *Nitellahyalina* var. *hyalina f. brachyactis* from Jammu. Comparative cytology and cytomorphology of *Nitellahyalina* var. *hyalina f. indica* has been investigated by Pundhir, *et al.*, (1993) from Rajasthan and cytology of *Charavulgaris* var. *imperfectaf. dissoluta* (for the first time) from Uttar Pradesh.

Pundhir, *et al.*, (1995 and 1999) have studied the cytology and cytotaxonomy of *Nitelladuthieae* from U.P. and *Lycnothamnus barbatus f. mahmoodia* forma. nov. from Jammu. Pundhir and Chauhan (1999) have studied the chromosome number of *Charasocotrensis f. socotrensis* R.D.W.

Euploid chromosome number, n = 14, 28, 35, 42, 49, 56, 63 and 70 have been reported for *Chara*. For *Nitella*, so far known are n = 6, 9, 12, 15, 18, 21, 24, 27, 33, 36 and 48. *Lycnothamnus* and *Lamprothamnium* possess n = 14, 28, 35, 63 and n = 14 and 42 respectively and Tolypella n = 6, 11, 15 and 20. Besides, anecuploid numbers n = 8, 26, 37 and 48 have also been reported in *Chara. Nitellopsis* is yet to be karyologically investigated in India.

II. SEX CHROMOSOMES

In most of the species of Charophyta, chromosomes have been studied from antheridial filaments of globules. This study is equal to meiotic studies in flowering plants. In dioecious members of animals and plants, sex chromosomes were identified from male and female reproductive cells. They are X and Y chromosomes respectively. But, so far in the antherical filaments of Charophyta, no sex chromosomes have been observed, according to Pundhir (1998). Therefore, he suggested that sex differentiation in dioecious and monecious races is controlled by genes rather than specific sex chromosomes. But, this suggestion may not be correct. In '*Characorallina*' n = 14 chromosomes have been observed in both monoecious and dioecious taxa. In both cases, the chromosomes were observed from antheridial filament cells of globules. But in dioecious taxa, chromosomes were not seen from Nucule cells. Therefore, sex chromosomes may be studied from root tip or shoot-tip mitosis so that we see the chromosomes of both male and female sexes separately. Then, if we make Karyotype analysis of the chromosomes of both sexual cells, we may better come to a conclusion, whether sex is expressed by gene or sex chromosome in Charophyta.

III. CYTOLOGICAL ABERRATIONS

In Charophyta the cytological studies from antheridial filaments of globules, is called meiosis and not mitosis. This antheridial filaments squashes are equal to pollen smear of flowering plants. Meiotic aberrations like grouping chromosomes, cells with micronuclei, lagging chromosomes and pseudobridges at anaphase were observed by Chatterjee (1979a) in *C. fibrosa* var. *fibrosaf. longicorollata* and he suggested its hybrid constitution. Subramanian (1983) has observed meiotic aberrations in *C. corallina* (Anaphasic laggards and precocious movements of chromosomes at metaphase side view), *Lycnothamnus barbatus f. barbatus* (Micro nuclei formations) and *Nitella pseudoflabellata* var. *imperialis f. wattii* (anaphasic laggard). Out of 26 taxa of Charophyta studied, 4 taxa have meiotic aberrations as shown by the author (Figure 1, Table 1). Therefore intensive studies in chromosomal aberrations are necessary. If we make pretreatment to young globules, before meiosis, the pretreatment chemicals will break the spindle fibres and spread the chromosomes inside the cell. Then it is not possible to see the meiotic aberrations inside the cells. Therefore, to get spreading of chromosomes pre-treatment is essential but to get the meiotic aberrations, we should not give pre-treatment before squashing the young globules. That is why, even though there may be large number of hybrids in Charophyta, the scientists could not see the meiotic aberrations and suggest the hybrid origin of them. They were satisfied only in the polyploidy nature of the species, due to increase in basic chromosome numbers.

All the studies on squashing the young globules of Charophyta are only meiotic studies and we should not treat them as mitotic studies. In flowering plants, the chromosomes are in pairs called bivalents in meiotic studies. Therefore, it is doubtful whether in Charophyta, the so called chromosomes reported by various scientists so far are typical chromosomes or bivalents, as only meiotic studies in Charophyta have been followed by various scientists of the world from globules, the male sex organ.

IV. CYTOTAXONOMICAL STUDIES

Cytotaxonomical studies have been made in any group of plants on the basis of chromosome number, size and morphology.

Based on cytotaxonomical studies, Sarma (1967) has divided the species of *Chara* into 1) *Chara vulgaris* complex, 2) *C. globularis* complex, 3) *C. setosa* complex, 4) *C.braunii* complex, 5) *C. corallina* complex and 6) *C. fibrosa* complex.

Similarly, Sarma (1967) has again formed distinct complexes among the species of *Nitella*. Accordingly, be recognized, 1) *Nitellaflexilis* complex, 2) *N. mirabilis* complex, 3) *N. stuartii* complex, 4) *N. dualis* complex, 5) *N. furcata* complex, 6) *N. tenuissima* complex, 7) *N. pseudoflabellata* complex, 8) *N. gracilis* complex, 9) *N. hyalina* complex and 10) *N. translucens* complex. Whatever species studied cytologically until that time were added together in the respective complexes.

V. CYTOGEOGRAPHICAL STUDIES

Cytogeographical distribution of Indian Charophytes has been discussed by Khan and Sarma (1981-82). They represent Ca. 33% of their world flora in India. Of these, eighteen taxa are endemic shared by *Chara*, *Nitella*, *Lychnothamnus* and *Tolypella*. Eight taxa were recognized as cosmopolitan, 25 taxa as sub-cosmopolitan and the rest were distributed as African, American, European and Pacific flora. Distribution of n = 14 and 28 chromosome numbers in *Chara* and n = 18 in *Nitella* in most of the investigated states, showed them more successful adaptive genomes with few exceptions. Next to these chromosome numbers were found n = 42 in *Chara* and n = 9 in *Nitella* distributed in 9 and 8 states of India respectively. Higher frequency of distribution and occurrence of polyploidy races in all the states in Indian Charophyta indicates the suitability of Indian habitats for the growth of the Charophytes.

VI. ORIGINAND EVOLUTION OF CHAROPHYTES

The origin and evolution of Charophytes have been discussed by many workers (Fritsch, 1935 Womersley and Ophel, 1947, Wood 1952, Chadefaud, 1960, Desikachary and Sundaralingam, 1962, Wood and Imahori, 1965). Fossils afford no help in this (Fritsch, 1935). The existence of undoubted fossil species of *Nitella* is not certain, although their presence in quarternary deposits would not be surprising (Horn and Rantzien, 1957). While discussing the affinities and interrelationships of Charophytes, Desikachary and Sundaralingam (1962) advocated Chaetophoralean origin of Charophytes. They hold the possibility of *Chara* being relatively primitive and *Nitella* derived. Further that ecorticated *Chara* are derived from corticated ones, whether haplostichous or triplostichous species are primitive is uncertain. It was later supported by Pal, *et al.*, (1962). Wood and Imahori (1965) considered *Nitella* more advanced than *Chara*. Among *Chara*, ecorticated forms are more evolved than corticated ones. Similarly, in *Nitella*, arthrodactylous forms have been considered more primitive than anarthrodactylous forms. Kasaki (1964) has considered that arthrodactylous forms are derived from anarthrodactylous forms. As yet no such genus has been discovered which may provide a link between *Chareae* and *Nitelleae*, the concept of parallel evolution of both the groups from a common ancestor, put forward by Womersley and Ophel (1947), Desikachary and Sundaralingam (1962), Wood and Imahori (1965) and Guerlesquin (1967) seem to be acceptable.

All these theories discussed on evolution and inter-relationships of this group were based on gross morphological Characters. A cytological approach to phylogeny interrelationships and evolution in Charophyta was made by Sarma, *et al.*, (1970). They have propounded that (i) tribe *Nitelleae* is primitive than tribe *Chareae*, (ii) arthrodactylous *Nitella* are advanced over anarthrodactylous *Nitella*, (iii) *Lychnothamnus* and *Lamprothamnium* are primitive forms, (iv) *Nitellopsis* segregated quite early from *Chara*. Further they have refuted the view of Chaetophoralean origin of Charophytes advocated by Desikachary and Sundaralingam (1962) on the basis of comparative assessment of nuclear cytology of Chaetophorales and *Charales*. Stebbins (1971) propounded that members with lowest chromosome numbers are more primitive. On this basis, it appears that *Nitelleae* are more primitive than *Chareae* as lowest chromosome number n = 6 are known in the species of *Nitelleae*, while in *Chareae* the lowest chromosome number so far recorded is n = 14.

Several phylogenetic schemes have been erected by diverse workers to show the evolution and interrelationships of this group on cytological grounds. Kasaki (1964) proposed that reduction in basic chromosome numbers has played a considerable role in the origin of all the taxa. Guerlesquin (1967) suggested an evolutionary sequence based on reduction in basic chromosome number for the evolution of *Nitelleae* and reduction cum-addition in the basic chromosome number for the evolution of *Chareae*. Origin of both the tribes from a common unknown ancestor and their parallel evolution; following the suggestion of Wood and Imahori (1965) and their further complicated segregation evidenced through emergence of different chromosome numbers was proposed by Sarma, *et al.*, (1970) in their tentative scheme of evolution of Charophytes. Sawa (1974) followed by Bhatnagar (1988c) presented hypothetical schemes through ancestral unknown stock X = 3 and subsequent parallel evolution of tribe *Chareae*, genera *Nitella* and *Tolypella*. More recently, Pal and Chatterjee (1986a, b) proposed a phylogenetic scheme based on feulgen's spectrophotometric estimation of DNA in relation with nucleus volume, chromatin length, DNA. content and chromosome number in the different species of Indian Charophytes. Their scheme is almost similar to that of Sarma, *et al.*, (1970) with a few deviations.

Some of the workers made karyotypic analysis in different taxa (Khan and Sarma (1967a, b), Ramjee and Sarma (1971), Chatterjee (1976a, b), Bhatnagar and Sharma (1985), Ray and Chatterjee (1988a, b), Pundhir and Gautam (1993, 1994a), Pundhir, *et al.*, (1993), Pundhir and Vidyavati(1994a, b). These analyses showed that each taxon is characterised by distinct chromosomal features. Marked differences in chromosomal morphology was noticed in the taxa having same chromosome numbers suggesting positive role of chromosomal rearrangements and hybridization in evolution and speciation in Charophytes. The karyotypic differences have also reflected in their morphological variations.

Certain conclusions have been drawn on the basis of observations of the above mentioned workers. These are given below:

a. Mitotic divisions in spermatogenous cells of Charophyte species followed a standard pattern of nuclear division as found in higher plants.

b. The discontinuous staining of chromosomes during prophase providing them beaded appearance were found to be the constant feature in all the taxa.

c. Association of chromocentres with nuclei were found to be a constant feature but here it has to be pointed out that the presence of extra-nucleolar bodies scattering somewhere in nucleoplasm tender the confusion to determine exact position and count of chromocentres.

d. No centriole or any other stained granules or bodies likely to resemble with it have been observed.

e. Basic chromosome numbers proposed as X = 7 for the species of tribe *Chareae* and X = 3 for the species of *Nitella*, are accepted by almost all the Charologists.

f. Subdivision of Charophyta into two tribes *Chareae* and *Nitelleae* has been justified on cytological grounds as tribe *Chareae* represented by basic chromosome number X = 7 and tribe *Nitelleae* represents X = 3 with the exception of genus *Tolypella*. Further subdivision of these tribes on cytological basis could not be possible.

g. Dioecious and monoecious conditions in Charophytes are independent and are not associated with the doubling of genomic complements of each other as supposed by Wood and Imahori (1965).

h. No perfect correlation was observed between chromosome numbers and chromosomes as a whole, but in the taxa of individual complexes some correlation exists to certain extent as proposed by Khan and Sarma (1967c) that smaller chromosomes are associated with higher numbers while larger chromosomes are associated with lower chromosome numbers, viz., the taxa of *Charaglobularis* complex.

i. No correlation between polyploidy and sexuality seems to exist in Charophytes.

j. Most of the Charophytes possess asymmetrical karyotypes irrespective of their morphological level of advancement.

k. Based on level of ploidy, tribe *Nitelleae* seems to be comparatively primitive than tribe *Chareae.*

l. Role of hybridization in inducing polyploidy among Charophytes widely supported.

m. Euploids in contrast to the aneuploids show wider range of distribution in India and in rest of the world.

n. Prevalency of low chromosome numbers and distinct series of euploidy in all the Indian genera tends the supposition of being Bihar, Uttar Pradesh and partially Madhya Pradesh as primary centre for the origin of world Charophycean flora and their subsequent migration to other continents.

o. Although existence of aneuploidy cannot be ruled out but Charophyte species showed highly complex state of euploidy.

p. It seems reasonable to conclude that polyploidy, hybridization and chromosomal rearrangements have a key role in the evolution and speciation in this group.

The above conclusions have been drawn on the basis of limited taxa studied. The areas of north-western region, extreme eastern region and central-western regions of India are still totally unexplored. Further intensive researches are required particularly on the hybridization, meiotic cycle and cytosystematics to unravel the problems of its relative fixity and obscured phylogenetical origin.

Chart - 1

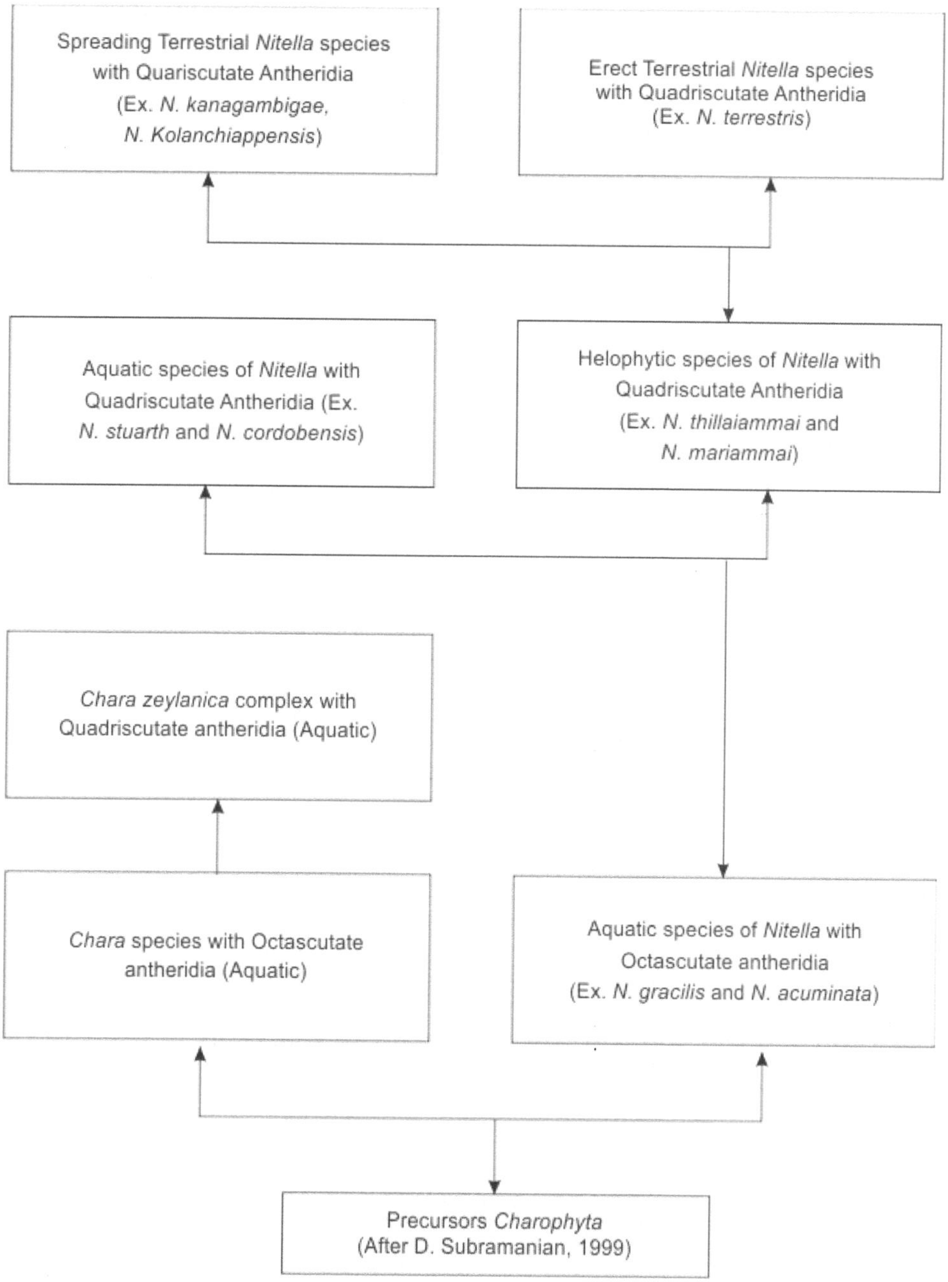
Spreading Terrestrial *Nitella* species with Quariscutate Antheridia (Ex. *N. kanagambigae, N. Kolanchiappensis*)
Erect Terrestrial *Nitella* species with Quadriscutate Antheridia (Ex. *N. terrestris*)
Aquatic species of *Nitella* with Quadriscutate Antheridia (Ex. *N. stuarth* and *N. cordobensis*)
Helophytic species of *Nitella* with Quadriscutate Antheridia (Ex. *N. thillaiammai* and *N. mariammai*)
Chara zeylanica complex with Quadriscutate antheridia (Aquatic)
Chara species with Octascutate antheridia (Aquatic)
Aquatic species of *Nitella* with Octascutate antheridia (Ex. *N. gracilis* and *N. acuminata*)
Precursors *Charophyta* (After D. Subramanian, 1999)

Chart 2 Hypothetical evolutionary sequence of the *Characeae* (after wood and Imahori, 1965)

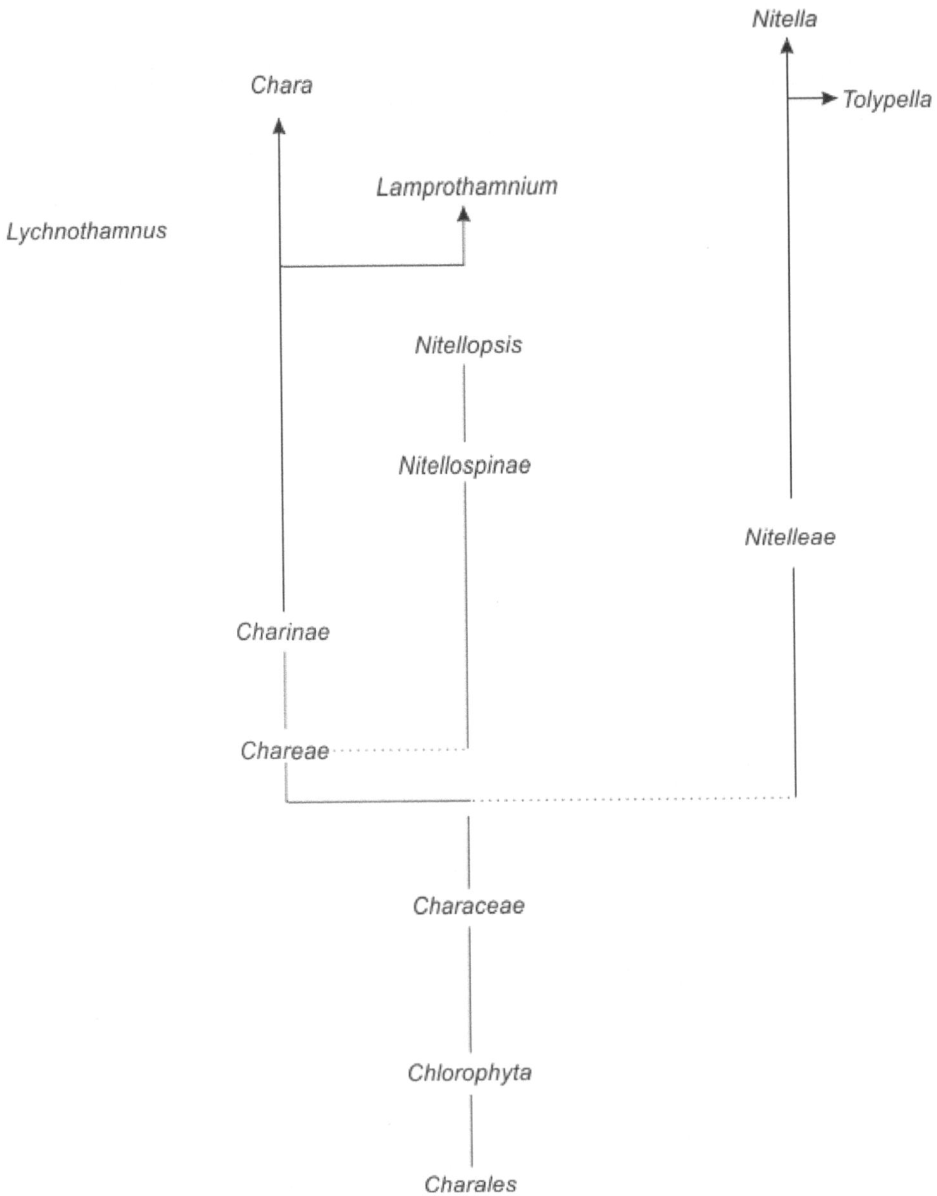

Chart 3 Simplified scheme of evolutionary sequence in *Charophytes* (Adopted from Guerlesquin, 1970)

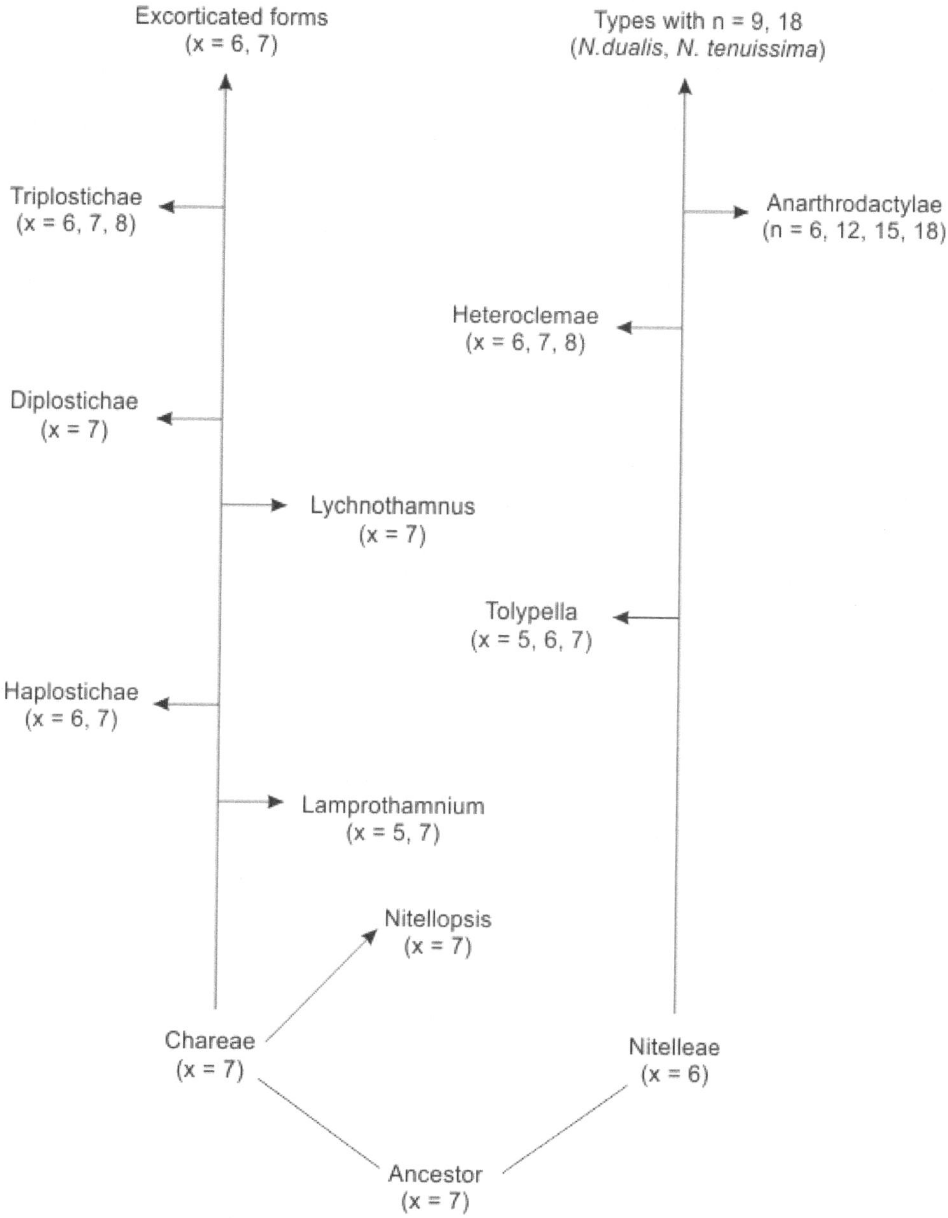

Chart 4 Evolutionary sequence in *Charophyta* based on cytological findings by Y.S.R.K. Sarma, M. Khan and Ramjee, 1970.

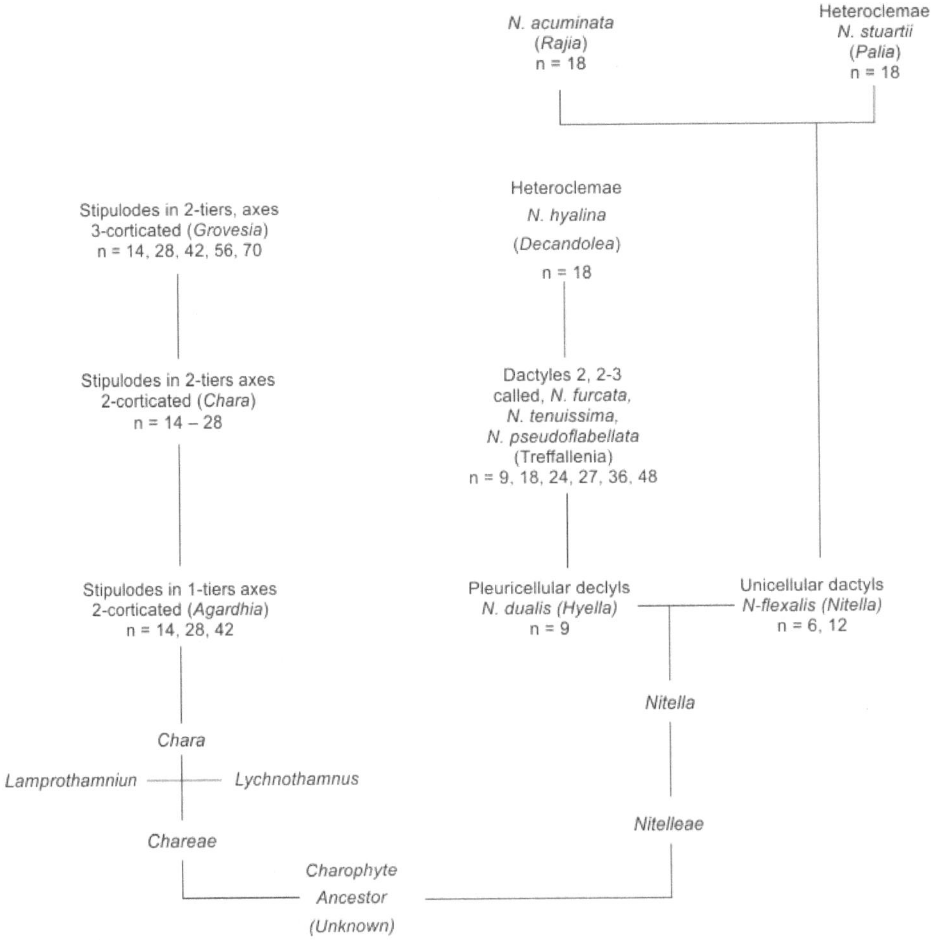

KARYOLOGY OF INDIAN CHAROPHYTA CYTOTAXONOMICAL STUDIES OF CHAROPHYTA FROM TAMIL NADU

Observations on interphase nuclei and meiosis have been made on young antheridial filaments in 25 taxa of charophytes from Tamilnadu. The chromosomes from the cells of antheridial filaments of *Chara, Nitella, Lychnothamnus* are illustrated in Fig. 1.

Precocious movements of chromosomes have been observed rarely in *Chara-corallina* (Fig.6). Anaphasic laggards have been noted in *C. corallina* (Fig.5), *Nitella furcata* subsp. *furcata* (Figs.21 and 22) and *N. pseudoflabellata* var. *imperialis f. wattii* (Fig. 34). Although they are rare in occurrence, nevertheless, they are indicative of polyploidy nature of the cells in *Chara, Nitella* and *Lychnothamnus* studied here.

Table – 1 shows the species studied, present reports of chromosome numbers and the previous reports in the respective species, authors and years. First records of chromosome numbers have been made in 4 taxa namely *N. pseudoflabellata* var. *imperialis f. wattii, N. gracilis* var. *athanurensis f. delicatula, N. gracilis* var. *athanurensis f. nana* and *N. gracilis* var. *athanurensis f. flexuosa*. Deviant reports of chromosome numbers as against the earlier reports have been made in *C. corallina f. corallina. C. braunii, C. fibrosa f. fibrosa. N. furcata* subsp. *mucronata. f. oligospira, N. furcata.* subsp. *flagelliformis. f. patula, N. acuminata f. acuminata, N. hyalina* and *Lychnothamnus barbatus f. barbatus* (Table – 1.).

Table – 1

Species studied	Present reports (n)	Previous reports (n), authors and years
Chara zeylanica f. elegans	28	Sarma and Khan (1967) = n = 28
C. zeylanica. var. *diaphana*	28	Sarma and Khan (1967) = n = 28
C. zeylanica. f. zeylanica	28	Sarma and Khan (1967) = n = 28
C. corallina. f. corallina	49+	Khan and Sarma (1967) = n = 42; Ramjee and Bhatnagar (1978) = ; n = 42
C. braunii	42+	Khan and Sarma (1957) = n = 14; Ramjee and Bhatnagar (1978) = n = 14
C. fibrosa. f. fibrosa	28+	Khan and Sarma (1967) = n = w = 14
C. fibrosa. f. erythrogyna	14	Khan and Sarma (1967) = n = w = 14
C. fibrosa	14	Hotchkiss (1965) = n = 28, 42; Khan and Sarma (1967) = n = 14
C. fibrosa. var. *hydropitys*	14	Khan and Sarma (1967) = n = 14
C. setosa f. setosa	28	Khan and Sarma (1967) = n = 28
Nitella furcata. subsp. *furcata*	18	Imahori and Kato (1961) = n = 24 Khan and Sarma (1967) = n = 18
N. furcata. subsp. *mucronata f. mucronata*	18	Gillet (1959) = n = 18; Khan and Sarma (1967) = n = 18
N. furcata. subsp. *mucronata f. oligospira*	36+	Khan and Sarma (1967) = n = 18

Species studied	Present reports (n)	Previous reports (n), authors and years
N. furcata. subsp. *mucronata f. flagellifera*	18	Khan and Sarma (1967) >> n = 18
N. furcata. subsp. *flagelliformis f. patula*	36+	Khan and Sarma (1967) = n = 18
N. furcata. subsp. *flagellifera*	18	Khan and Sarma (1967) = n = 18
N. acuminata. f. acuminata	12+	Hotchkiss (1958) and Imahori and Kato (1961) = n = 18; Ramjee and Bhatnagar (1978) = n = 18
N. pseudoflabellata var. *elegans*	18	Khan and Sarma (1967) = n = 18
N. tenuissima f. bengalensis	18	Khan and Sarma (1967) = n = 18
N. hyalina	12+	Ramjee and Bhatnagar (1978) = n = 18
N. translucens	18	Khan and Sarma (1967) = n = 18
N. pseudoflabellata var. *imperialis f. wattii*	18*	----
N. gracilis var. *athanurensis f. delicatula*	18*	-------
N. gracilis var. *athanurensis f. nana*	18*	----------
N. gracilis var. *athanurensis f. flexuosa*	18*	--------------
Lychnothamnus barbatus f. barbatus	60+	Hotchkiss (1963) = n = 28; Khan and Sarma (1967) = n = 28.

* = First record of chromosome numbers.

\+ = Diviant report of chromosome numbers.

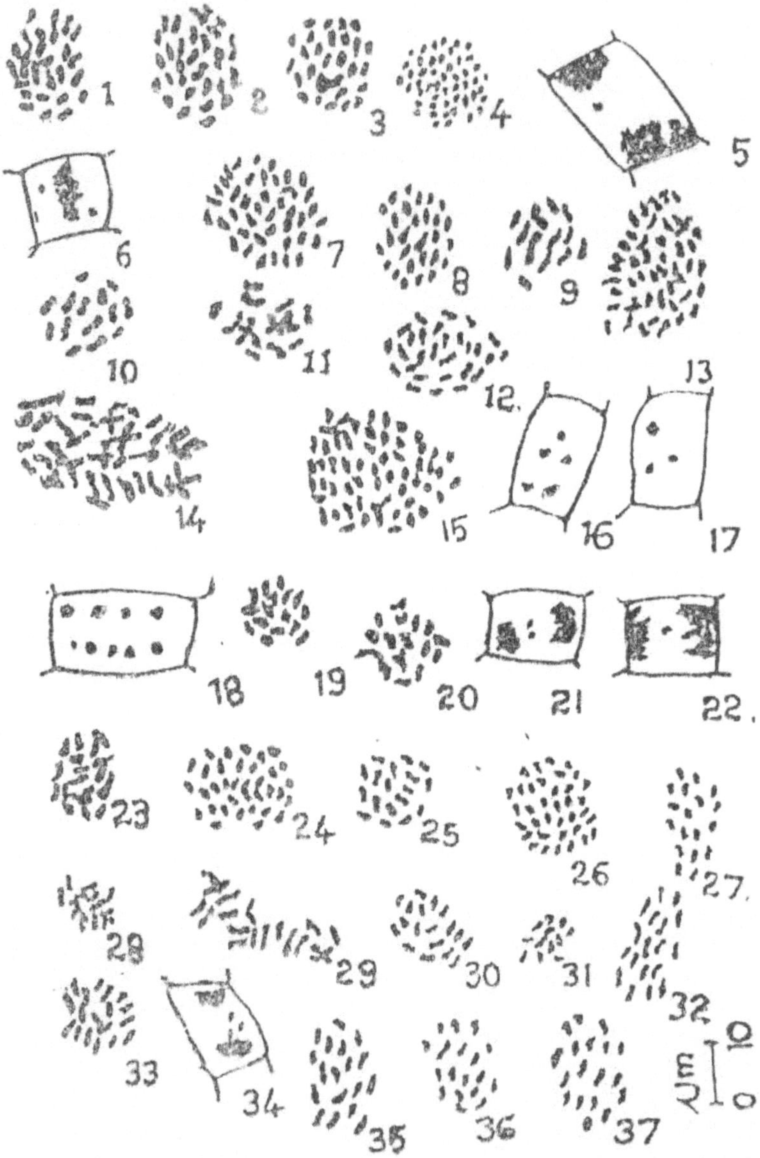

Figure– I 1. *Chara zelanica. f. elegans* n = 28; 2. *C. zeylanica.* var. *diaphana* n = 28; 3. *C. zeylanica.* var. *zeylanica* n = 28; 4. *C. corallina,* n = 49, 5. and 6. *C. corallina,* Anaphasic laggard and precocious movements. 7. *C. braunii* n = 42 ; 8. *C. fibrosa f. fibrosa* n = 28

; 9. *C. fibrosa f. erythrogyna* n = 14; 10. *C. fibrosa,*n = 14; 11. *C. fibrosa* var. *hydropitys* n = 14; 12. *C. setosa f. setosa*n = 28; 13. to 18. *Lychnothamnus barbatus f. barbatus*; 19 to 22. *N. furcata* subsp. *furcata* n = 18; 23. *N. furcata* subsp. *mucronata f. mucronata* n = 18; 24. *N. furcata* subsp. *mucronata f. oligospira*n = 36; 25. *N. furcata* subsp. *mucronata f. dictyosperma*; 26. *N. furcata* subsp. *flagelliformis f. patula* n = 36; 27. *N. furcata* subsp. *flagellifera,* n = 18; 28. *N. acuminata f. acuminata* n = 12; 29. *N. pseudoflabellata* var. *elegans* n = 18; 30. *N. tenuissima f. bengalensis* n = 18; 31. *N. hyalina* n = 12; 32. *N. translucens* n = 18; 33. and 34. *N. pseudoflabellata* var. *imperialis f. wattii;* 35. *N. gracilis.* var. *athanurensis f. delicatula* n = 18; 36. *N. gracilis.* var. *athanurensis f. nana* n = 18 and 37. *N. gracilis* var. *athanurensis f. flexuosa* n = 18.

It is concluded that polyploidy plays an important role in speciation in *chara* and *Nitella, Lychnothamnus barbants* with n = 60 chromosomes is a higher polyploidy. In the smaller size and more number of chromosomes, it is more related to *chara corallina* and *C. braunii.*

III

CHEMOTAXONOMICAL STUDIES

Eventhough cytotaxonomical studies on Charophyta have been made all over the world, biochemical or chemotaxonomical studies or medical aspects of this group of plants have not at all been studied. In this chapter, a preliminary approach has been made on nature of free amino acids, protein bound aminoacids and chlorophyll a, b and total chlorophyll content in some of the species. The results of these studies had been published in proceedings of all India applied phycological congress. I. Kanpur (pp. 107-112) and advances in applied phycology. II. Kanpur (pp. 305-206) by the author in 1982 and 1989.

The details of these studies are presented below.

CHEMOTAXONOMICAL STUDIES OF CHAROPHYTA FROM TAMIL NADU BY D. SUBRAMANIAN

1. Study of chlorophyll a, b and total chlorophyll content and protein bound amino acids.

A large number of species of Charophyta have been studied morphologically (Pal, *et al.*, 1963; Sundaralingam, 1959; Sundaralingam, *et al.*, 1959 and Wood and Imahori, 1965) and cytologically by a large number of authors in India. But, there is no information about the chemotaxonomical studies of Indian Charophyta. The present investigation is a part of such studies to fill up this gap. At present some of the available taxa from Tamil Nadu have been included for chemotaxonomical studies with special reference to protein bound amino acids and chlorophyll.

Materials and Methods

Lychnothamnus barbatus f. barbatus (Meyen), Leon and *Nitellopsis obtusa* (Desv.), Groves were obtained from Madras. *Nitella acuminata* Br. Ex. Wallm was collected from Ootacamund of Western Ghats. All other species studied were collected from Virudhachalam and Chidambaram Taluks of South Arcot District of Tamil Nadu. All species were grown in well aerated glass jars in Botanic Garden under identical conditions.

Two dimensional chromatograms were developed in triplicate at room temperature (27°C) in Whatman No. 1 chromatogram paper (29.5 cm x 23.0 cm) from extracts of fresh plant parts of the same age with sex organs. Half gram of the material was ground with 10 ml of 95% alcohol : Conc. HCl (99:1) mixture using glass mortar and pestle. Then the solution was kept at 60°C for 30 mts. After adding 30 ml of chloroform and shaken thoroughly, the mixture was kept in dark overnight. Afterwards, the supernatant was separated using a separating funnel and it was centrifuged to remove the debris and used for spotting. The solvent for the first run was a mixture of n-butanol : acetic acid: water (67 : 16.5: 16.5 v/v) and for the second run a mixture of phenol : water (85 : 15 v/v). The amino acids were detected by spraying chromatogram paper with 0.2% ninhydrin dissolved in 95% acetone followed by a brief heating at 105°C (5 mts). By studying colour of the spots and calculating their R_f values, the aminoacids were identified. Authentic samples of 22 amino acids were run in triplicate at room temperature (27°C) already in the laboratory. These known samples were used to identify different amino acids present in each species. Half gram of fresh material in each species was ground in glass mortar and pestle with 10 ml of 80% acetone. The extract was centrifuged and supernatant was separated away by adding acetone. Transmittance and absorbance values were estimated at 645 nm and 665 nm at room temperature (27°C) in spectronic 20. The amount of chlorophyll a, chlorophyll b and total chlorophyll in each species studied was then calculated and results tabulated (Table – 2) following the method of Arnon (1949).

Observations

As many as 15 amino acids were observed in the species of *Chara*, *Nitella*, *Lychnothamnus* and *Nitellopsis*. The minimum number of spots noted is 7 in *Charasetosa* and maximum number of spots 11 is observed in *Nitellopsis obtusa*. As far as the number and type of amino acids are concerned (Table – 4). *Nitellopsisobtusa* is related more to the species of *Nitella* and *Lychnothamnus barbatus* than to the species of *Chara*.

Aspartic acid, glutamic acid, arginine and lysine are present in all the species studied and cysteine, threonine, phenyl alanine, asparagine and glycine are present in most of the species.

Table – 2

Species	Chlorophyll a in gm / ml	Chlorophyll b in gm / ml	Total chlorophyll in gm / ml
Nitella acuminata	0.5565300	0.2297200	0.7862500
N. hyalina	0.4491830	0.3913700	0.8405530
N. pseudoflabellata var. *elegans*	0.3297400	0.2915900	0.6216600
N. furcata	0.3676720	0.2432355	0.5109460
Nitellopsis obtusa	0.3685240	0.2765480	0.6450720
Lychnothamnus barbatus f. barbatus	0.3576180	0.1522920	0.5099100
Chara corallina	0.2059750	0.1482320	0.3542070
C. fibrosa var. *hydropitys*	0.3376065	0.1276750	0.4652815
C. fibrosa var. *gymnopitys*	0.3450070	0.1441790	0.4897180
C. fibrosa f. fibrosa	0.3362760	0.0859530	0.4225370
C. setosa	0.3136230	0.1833670	0.4969900
C. zeylanica var. *macilenta*	0.2869795	0.1980640	0.4850435

Alanine is present only in *Nitella* species which distinguished *Nitella* from *Chara*, *Lychnothamnus* and *Nitellopsis*. Due to presence of particular amino acids, *Lychnothamnus* is related to *Charasetosa* and is related to *C. zeylanica* var. *macilenta* and *C. corallina* by presence of same number of amino acids.

The total chlorophyll content in gm / ml ranges from 0.48832 to 0.786250 in *Nitella* species. The maximum value is observed in *N. acuminata* and the minimum value in *N. hyalina* (Table – 2). *Nitellopsis obtusa* is related more to the species of *Nitella*, as evidenced by the total chlorophyll content. In *Chara* species, the total chlorophyll values range from 0.354207 gm / ml to 0.496990 gm / ml. The maximum value is in *C. setosa* and the minimum value in *C. corallina* and *Lychnothamnus barbatus f. barbatus* is in between *Chara* and *Nitella* (Table – 2) for its chlorophyll contents.

Discussion

Since the publication of the book "Biochemical Systematics" by Alston and Turner (1960), emphasis has been laid on comparative biochemical data for supplementing morphological characteristics of different taxa in order to assess the affinities of species (Buzatti-Traverso, 1969).

Table – 3

Name of amino acids	Colour	Rf value in 1st solvent system R_f : (x 100) BAW	Rf value in 2nd solvent system R_f : (x 100) in Phenol : H_2O
Aspartic acid	Bluish violet	16.7	5.1
Lysine	Pink	6.3	5.1
Glycine	Red violet	22.9	15.5
Tyrosine	Pinkish grey	30.0	36.0
Glutamic acid	Pinkish violet	16.7	46.2
Arginine	Pinkish violet	16.6	51.3
Leucine	Violet	54.2	72.3
Valine	Violet	66.6	72.3
Iso-leucine	Violet	54.2	61.5
Alanine	Violet	41.7	64.1
Serine	Violet	20.2	30.5
Asparagine	Orange violet	33.3	40.6
Threonine	Violet	29.2	60.2
Cystine	Violet	8.3	25.8
Phenyl alanine	Grey violet	45.8	50.6

The works of Alston and Irwin (1961) and Alston and Turner (1962) have shown that there are variations in chromatographic pattern of amino acids in different species of same genus. The same principle has been successfully followed in the author's laboratory [Subramanian (1978), Subramanian and Krishnamurthy (1978) and Krishnamurthy and Subramanian (1978)], to solve the problems of interrelationships among the species of the same genus and among the species and genera of a particular family of Angiosperms.

In general, the species of Charophyta are chemically more specialized. As far as the amino acids are concerned there are 15 amino acids occurring in various combinations (Table – 4).

Table – 4

Species	No. of amino acids	Name of the amino acids
Nitella acuminata	10	ASP ; GLU; ARG; LEU; ALA; PHE-ALA; ASPAR; LYS; GLY; SER
N. hyalina	10	ASP L GLU; ARG; LEU; ALA; PHE-ALA; ISO-LEU; GLY; LYS; ASPAR
N. pseudoflabellata	10	ASP; TYR; GYS; GLU; ARG; ALA; THR; ASPAR; LYS; GLY
N. furcata	11	ASP; GYS; THR; GLU; ARG; ALA; ASPAR; LYS; ISO-LEU; LEU;
Nitellopsis obtuse	11	ASP; GYS; GLU; ARG; LYS; GLY; ASPAR; LEU; PHE; ALA; SER;
Lychnothamnus barbatus f. barbatus	9	ASP; GYS; GLU; ARG; THR; PHE-ALA; ASPAR; LYS; GLY
Chara coralina	9	SER; ASP; GLU; THR; PHE-ALA; ASPAR; LYS; GLY; GYS
C. fibrosa var. *hydropitys*	8	GYS; GLU; LYS; ARG; ASPAR; SER; VAL; TYR
C. fibrosa var. *gymnopitys*	9	ASP; GYS; GLU; LYS; ARG; ASPAR; SER; VAL; TYR
C. fibrosa f. fibrosa	7	ASP; GYS; GLU; LYS; ARG; ASPAR; VAL
C. setosa	7	ASP; GYS; GLU; LYS; ARG; ASPAR; SER
C. zeylanica var. *macilenta*	9	ASP; CYS; ASPAR; GLU; ARG; VAL; THR; PHE-ALA; LYS

Legend : SER – Serine ; ASP – Aspartic Acid ; GLU – Glutamic Acid ; ARG – Arginine; CYS – Cystine ; LEU – Leucine; ALA = Alanine ; PHE-ALA – Phenyl Alanine; ASPAR – Asparagine ; HIS – Histidine ; VAL – Valine; ISO-LEU – Isoleucine ; THR – Threonine ; LYS – Lysine; TYR – Tyrosine.

The chemical specialization is as much as found in advanced Angiospermic taxa and families like Compositae (Rani and Subramanian, 1979), Rubiaceae (Selvaraj and Subramanian, 1980), Acanthaceae (Subramanian and Krishnamurthy, 1978) and *Leersia* and *Oryza* (Krishnamurthy and Subramanian, 1978).

Among the species studied, the species of *Nitella* and *Nitellopsis* are chemically more specialized as they have larger number of amino acids, as compared to those of *Chara* and *Lychnothamnus*. The morphological characteristics again show the same conclusion. Besides, the chlorophyll estimation in these species reveals more or less a similar conclusion.

2. Study of free amino acids by D. Subramanian

In 4 species of *Chara*, 6 species of *Nitella*, *Lychnothamnus barbatus f. barbatus* and *Nitellopsis obtuse*, Chemotaxonomical studies with special reference to free amino acids have been made in order to understand the interrelationship among them as this sort of knowledge does not exist so far in this group of plants.

Fresh thallus with sex organs in each species was dried in oven at 100°C for 24 hrs. and 0.5 gm powder in each was taken. 5 ml of alcohol: NHC1 was taken in a pencillin bottle in which the dried sample was kept for 24 hrs. in dark at room temperature (27°C). Then, the extract of the sample was spotted on Whatman No. 1 Chromatogram paper (29.5 cm x 23.0 cm).

Two dimensional chromatograms were developed in triplicate at room temperature to avoid artifacts. The solvent for the first run was a mixture of tertiary butyl alcohol; acetic acid; water (3:1:1). Then after drying, second run was given with solvent acetic acid : water (85:15). After drying, the fluorescent spots were observed in the media of ultraviolet light and ammonia vapour. The number of colour of spots in each species were noted and the results given in the following table :

Table – 5

S. No.	Taxa	No. of spots	Colour of spots
1.	*Chara braunii* Gm.	4	BR. G; Pl. W; Pl. W; Y.G.
2.	C. corallina Willd.	6	BR.G; Ll. Pl. W; Ll. Y; G. BR; BR; Ll. BR.
3.	*C. zeylanica f. zeylanica* R.D.W.	5.	Ll. BR. G; D1. BR.G; BR; Ll. BR
4.	*C. fibrosa* R.D.W.	5	Ll. BR. G; D1. BR. G; BR. G; BR; Ll. BR.
5.	*Lychnothamnus barbatus f. barbatus* (Meyen) Leon.	3	Ll. BR. G; BR.G; BR.
6.	*Nitella furcata* subsp. *furcata* R.D.W.	7	Ll. BR.G; BR.G; BR; Ll. BR; Ll.G; BR; BL.G.
7.	*N. pseudoflabellata* var. *elegans* R.D.W.	6	Ll.G; BR.G; Ll.G; Dl.G; Ll.BR.G; Dl.BR.G;
8.	*N. gracilis* var. *athanurensis* D. Subramanian	5	BR.G; Bl.G; BL.G; Y.G.; BR.
9.	*N. acuminata f. acuminata* R.D.W.	6	BR.G.; BL.G; Y.G; Y; G; Ll.BR; BR.
10.	*N. translucensAgardh.*	5	Ll.BR.G; BR.G; Ll.Y.G; Ll.BR; Dl.BR.
11.	*N. hyaliana* (DC) Ag. R.D.W.	6	BR.G; BL. G; Y.G; Y.G; Ll. BR; Dl. BR
12.	*Nitellopsis obtusa* (Desv.) Groves	7	Dl. BR.G; BR.G; Dl. BR; BR; Ll.BR; BL.G; Ll. BL.G.

Legends for colours BR.G = Brownish green ; Ll.BR; G – Light brownish green; Dl.BR.G. – Diffuse brownish green ; Dl.BR – Diffuse brown; Ll.BR – Light brown; BR – Brown; Ll.G – Light green ; Dl.G. – Diffuse green ; Y.G. – Yellow green; Ll.Y.G. – Light yellowish green; BL.G. – Bluish green; Pl.W. Pinkish white; Ll.Pl.W – Light pinkish white.

The name of the species and the number of colour of free amino acids are furnished in Table – 5. *Lychnothamnus barbatus* has the minimum number of amino acids, that is, 3 and *Nitellopsis obtusa* and *N. furcata* the maximum number of 7 amino acids. Among the species of *Chara* studied, the number of amino acids ranges from 4 to 6. Likewise, the number of amino acids among the species of *Nitella* studied, ranges from 5 to 7. Therefore, the species of *Nitella* have more number of amino acids in general showing their chemically more specialised nature, when compared to the species of *Chara* studied. *Lychnothamnus barbatus* is more related to the species of *Chara*, whereas *Nitellopsis obtusa* to the species of *Nitella* based on the number and colour of amino acids as revealed by the present chemotaxonomical studies. The same conclusions have been made while these taxa have been studied with special reference to protein amino acids and chlorophyll a, b and total chlorophyll (Subramanian and Ganesan, 1983) showing relationship of *Nitellopsis obtusa* to the species *Nitella* and *Lychnothamnus barbatus* to the species of *Chara*.

3. Study of comparative estimation of DNA and RNA and flavonoid aglycones in *Chara* and *Nitella* by D. Subramanian

The present study deals with the comparative estimation of DNA and RNA in mg / 1 gram fresh material among the species of *Chara* and *Nitella* and flavonoid aglycones. This is to understand the interrelationship of the species available in Tamilnadu. The members of Charophyta may be usefully exploited in the future. Therefore, the knowledge of the biochemical aspects of these species is very important and so, the present investigation has been undertaken.

Materials and Methods The materials for the present investigation have been collected at Kolappakkam and Mavidanthal of Virudhachalam Taluk and Portonovo and its neighbourhood of Chidambaram Taluk of South Arcot District of Tamilnadu.

For the estimation of Nucleic acids, the extraction and estimation procedures are given below:

Extraction procedure 100 mg of fresh leaf material was macerated in a mortar and pestle with 10 ml of cold methanol. The homogenate was centrifuged at 2000 rpm for 15 minutes. The supernatant was discarded to remove the methanol soluble phosphorus. The residue was washed with 10 ml of cold 0.5 N perchloric acid (PCA) and centrifuged. The supernatant was discarded to remove acid soluble phosphorus. The residue was extracted with boiling ethanol: ether in the ratio of 3:1 (V/V). This was boiled in a waterbath (90°C) for 20 seconds and centrifuged. The supernatant was discarded to remove the lipid phosphorus. The residue contains nucleic acids-p. The residue was extracted with 10 ml of 0.5 N PAC in a water-bath at 90°C for about 15 minutes and centrifuged. The supernatant was saved and made up to 10 ml using 0.5 N PCA.

This extract was used for the test of RNA and DNA.

Estimation of RNA

0.5 ml of extract was taken in a test tube and 2 ml of the orcinol reagent was added. The test tubes were heated for 8 minutes on water bath maintained at 100°C. The test tubes were taken out and cooled under a running tap. Reagent blanks were also maintained. The absorbance was measured in spectrophotometer at 665 nm using an orcinol blank.

Reagents used

1. 0.5 N PCA
2. Orcinol reagent.

Preparation of Reagents

1. *0.5 N PCA*

8.33 ml of 60% PCA was dissolved in distilled water and made up to 100 ml.

2. *Orcinol Reagent*

1 gm of purified and recrystallized orcinol in 100 ml of conc. hydrochloric acid containing 0.5 gm of ferric-chloride added.

S. No.	Species studied according to Pal, *et al.,*	Species studied according to Wood and Imahori	Place of collection	Date of collection
7.	*Nitella patula.* J.Gr. & G.O.A.	*N. furcata* subsp. *flagelliferaf. patula* (J.Gr. & G.O.A.) R.D.W.	Portonovo	03.03.88
8.	*N. pseudoflabellata* var. *imperialis.* T.F.A.	*N. pseudoflabellata* var. *imperialis* T.F.A.	Kolapakkam	27.03.88
9.	*N. elegans* Pal.	*N. pseudoflabellata* var. *elegans* (Pal.) R.D.W.	Kolapakkam	27.03.88
10.	*N. acuminata*	*N. acuminata f. klankinshipii* (T.F.A.) R.D.W.	Ko. Athanur	27.03.88
11.	*N. flexilis* (L) Aq. Em.	*N. flexilis* (L) Aq. Em.	Mavidanthal	27.03.88
12.	*N. spiciformis* Her	*N. gracilis* subsq. *cloeostachys* var. *spiciformis* (Mer) R.D.W.	Mavidanthal	27.03.88
13.	*N. superba* Pal.	*N. dualis* var. *pulchella f. superba* (Pal.) R.D.W.	Mavidanthal	27.03.88
14.	*N. mucronata* (A.Br.) Miq.	*N. furcata* subsp. *mucronata* (A.Br.) R.D.W.	Portonovo	03.03.88

Table – 6 Estimation of flavonoid aglycones

S. No.	Species studied	No. of spots	Colour of spots
1.	*Chara zeylanica*	6	BR.G;Y.BR; BR. Dl.BR. Y; Ll.BR; Ll.GR.
2.	*C. hydropitys*	5	BL.G. Ll.Y.BR; BR; Ll.BR; Ll.GR;
3.	*C. gymnopitys*	5	Ll.BL.G; Y.BR; Ll.BR; BR.Y; Ll.BR;
4.	*C. brachypus*	6	BL.G; Y.BR; Ll.BR; BR: BR.GR; GR.
5.	*C. braunii*	5	BL.G; Y; Dl.Y; BR; Ll.BR.Y;
6.	*C. corollina*	5	GR;BR.Y; Ll.BR.Y; BR; Ll.Y.BR;
7.	*Nitella patula*	6	Ll.G.Y; BR.Y; BR; Ll.BR.Y; Ll.BR; Ll.BR.G.
8.	*N. pseudoflabellata*	6	G.Y; Ll.BR.Y; BR; BR.Y; Ll.BR; BR.G;
9.	*N. eleqans*	8	G.Y; Ll.G.Y; BR.Y; Ll.BR; BR; Ll.BR.G; BR.G;
10.	*N. flexilis*	7	Dl.G.Y; BR; BR.Y; Ll.BR.Y; Ll. BR; BR.G; Dl.BRG;
11.	*N. acuminata*	5	Ll.G.Y; Ll.BR; Ll.BR.Y; BR; BR.G;
12.	*N. gracilis*	7	G.Y; BR.G; BR.Y; Ll.BR.Y; Ll.G; Y.BR; Ll.BR.
13.	*N. superba*	5	Ll.G.Y; Ll.BR.G; BR.G; BR.Y; Ll.BR.Y;
14.	*N. mucronata*	6	Ll.G.Y; BR.G.Ll; BR.G.; BR.Y.Ll; BR.Y; Dl.BR.Y;

Legands for Colours BR.G = Brownish green; BR = Brown; DI. BR.Y = Diffuse brownish yellow; BL.G = Bluish green. LI.Y.BR = light yellowish brown; LI.BR = light brown; LI.GR = light green; LI.BL.G – light bluish green; Y.BR = yellowish brown; BR.YU =Brownish yellow; Y = yellow LI.BR = light brown; G.Y. = Greenish yewllow; Y.BR = yellowish brown.

Table – 7 Properties of Flavonoid aglycones detected on paper Chromatograms of charophytes

S. No.	Group / possible identity	Fluorescence under UV + NH_3	Rf I*	RF II*
1.	Flavonol / 3.9 methyl – Kaempferol	Brownish yellow (BR.Y; Ll.BR.Y)	86	91
2.	Flavone / Apigenin	Green (G; Ll.G)	75	85
3.	Flavone ./ lricetin	Greenish yellow (G.Y; Ll.G.Y)	70	59
4.	Flavonol / Ramnetin	Brownish green (BR.G; Ll.BR.G;)	94	71

Rf values are mean values of 3 replications expressed as Rf x 100

I* - Rf EAW – Et Ac – Ho AC – H_2O in 50:2.50 (V/V/V) upper phase

II* - Rf BAW – Bu OH OH – HO AC – H_2O in 4:1:5 (V/V/V) upper phase

Table – 8

S. No.	Species studied	Amount of DNA in mg/g material	Amount of RIM A in mg/q material
1.	*Chara zeylanica*	1.623	2.014
2.	*C. hydropitys*	1.032	1.813
3.	*C. gymnopitys*	0.875	1.357
4.	*C. brachypus*	1.204	1.875
5.	*C. braunii*	0.950	1.258
6.	*C. corollina*	0.759	1.195
7.	*Nitella patula*	1.756	2.226
8.	*N. pseudoflabellata*	1.753	2.056
9.	*N. elegans*	1.875	2.254
10.	*N. flexilis*	1.125	1.813
11.	*N. acuminata*	1.603	2.023
12.	*N. gracilis*	1.625	2.071

S. No.	Species studied	Amount of DNA in mg/g material	Amount of RIM A in mg/q material
13.	*N. superba*	1.050	2.053
14.	*N. mucronata*	1.125	1.758

IV

OBSERVATIONS AND DISCUSSIONS

1. ECOLOGICAL ADAPTATIONS OF CHAROPHYTA

Charophytes are more specialized and evolved plants in Algae. They are kept in equal status along with Bryophytes in evolutionary ladder or just below it (Leedale, 1974; Chapman, 1980 and Coleman, 1980).

Like flowering plants, Charophytes are capable of living both in fresh water and salt water conditions. They live in multifarious climatic conditions throughout the world. Most of the plants live in fresh water in a submerged but shallow water level. A few plants of *chara* live inside salt water. But, all the species of the only genus *Lambrothamnium* live only inside salt waters. The Charophytes have some adaptive mechanisms either to overcome the external disturbances or to establish their progenies for further generations during adverse situations.

I. ADAPTATIONS OF CHAROPHYTES IN SANDY SOIL AND SAND DUNES

In most of the soil types, the water will be stagnant for 2 or even 3 months, after heavy rainfalls during monsoon season every year. But in Periappattu and Puduchattiram of Cuddalore district, and Kannankulam of Nagapattinum district of Tamil Nadu there are sandy cultivated fields intermittent with sand dunes here and there. During rainy seasons, the farmers cultivate paddy only on the rain water source and there is no irrigation of water either by natural means or by motor pump sets. After heavy rainfall, the paddy fields are full of water. Inside paddy plants, there are Charophytes. After raining, the water level is fastly going down. At this period the normally growing Charophytes like, *Chara fibrosa* Ag ex. Bruez, *Nitella furcata* Ag. em *N. tenuissima* (Desv) Katz *N. pseudoflabellata* A. Br. and *N. flexilis* (L) Ag. quickly react themselves to reduce the vegetative growth and produce more number of fertile whorls. As a result, all upper whorls are converted into smaller and more condensed and shorter fertile whorls. If the *Nitella* species appear lately after the close of the raining, they have only one or two basal vegetative whorls and all other upper whorls become fertile ones. At this stage these plants appear to be the highly condensed, shorter and distinct plants (Plates 1a, 1b, 1c, 2a, 2b, and 2c) when compared to the respective normal plants present in some perennial small lakes or ditches by the side with full of water almost to ¾ of the year (Plates 1 and 2). If this process is repeated for several years, or if there is poor rainfall for 2 to 3 years consecutively, there may be the possibility of origin of new forms, or varieties or sub species or even species.

If we study the adaptive mechanisms of Charophytes after heavy rainfalls, for 4 to 5 consequetive weeks, we can clearly observe the changes in their further growth. They finish their life cycle within 20 days to 1 month. In *Chara*, the normal plants (Plate 3) have sex organs in the second and third nodes of branchlets. But water – stress affected plants (Plates 3a and 3b) will produce sex organs in all the nodes of t Nitella grandiformis from Madras he branchlets. All these plants produce more oospores than necessary so that further generations will be ensured with these seeds like oospores.

Important morphological features of the modified plants due to water stressing are given below.

1. ***Nitella furcata var. mucronata f. puduchattirensis* forma nov. D. Subramanian (2005), National Seminar on present scenario in plant science Research, AnnamalaiUniversity, Abs. No. 1.2, pp. 7&8. (Plate - 2b).**

Plant monoecious, dwarf and bushy, repeatedly branched, 5 to 6 cm in height; 7 primary rays at each node of axis, each primary ray into 5 secondary rays of unequal lengths; each secondary ray into 1 tertiary ray and 3 to 4 dactyls of unequal lengths; each tertiary ray into 2 to 3 shorter dactyls; dactyl 2 celled, the tip cell elongated, conical and narrow; oogonium brown, 8 spiral wall convolutions; oospore with spinous projections; oogonium 380 to 400 μm in length and 300 to 320 μm in breadth; antheridium orange, 350 to 360 μm in diameter; sex organs in the axils of first and second branchlets of the whorls.

Collection No. 208. Holotype at author's home laboratory. Isotype No. 70 at B.S.I., CBE, T.N.

2. *Nitella pseudoflabellata* var. *angelii* var. nov. D. Subramanian, 2005. National Seminar on Present Scenario in Plant Science Research, Abs. No. 1.2 : pp. 7&8. (Plate – 1a).

Plant monoecious, medium sized and few branched, to 14 to 16 cm high, axis slender, 320 to 350 μm thick; internodes longer than branchlets; separate sterile and fertile whorls upto the medium part of the plants and gradual change from sterile whorls to fertile whorls forming 3 to 4 clusters of fertile whorls one above the other at the apical parts of the plants. At each node along with one fertile whorl, 7 primary rays forming ½ of the length of the entire branchlets; each primary ray into 7 secondary rays and each secondary ray into 1 or 2 tertiary rays and 4 to 5 dactyls; each tertiary ray into 2 to 3 dactyls; sterile whorls also with sex organs; rarely, all the fertile whorls enclosed in mucous clouds. Only primary and secondary rays ending into dactyls in fertile whorls and the pattern of furcation of fertile whorls similar to sterile whorls; at the node of primary ray, antheridium present and at the node of the secondary ray oogonium present. The primary rays, secondary rays and dactyls curved and shorter in fertile whorls; both antheridia and oogonia stalked (2 to 3 cells); antheridium large, orange coloured, 450 to 480 μm in diameter; oogonium small and brown; 350 to 380 μm in length and 300 to 320 μm in breadth; oospore membrane reticulate; 9 spiral wall convolutions.

Collection No. 212, Holotype at author's home laboratory and Isotype No. 49 at B.S.I., CBE, T.N.

3. *Nitella pseudoflabellata* var. *pantherii* var. nov. D. Subramanian, 2005, National Seminar on Present Scenario in Plant Science Research, Abs. No. 1.2, pp. 7 & 8. (Plate - 1b).

Plant monoecious, medium sized to 9 to 11 cm high, rarely branched; 2 to 3 sterile whorls at the lower part of the plant; the axillary and terminal whorls of branchlets fertile without mucous clouds; sterile whorl into 6 primary rays, each into 5 secondary rays, each secondary into 2 to 3 dactyls and 1 tertiary ray ending into 2 to 3 dactyls. Fertile whorls comparatively smaller with lesser mucous, each into 6 primary rays, each primary ray into 6 dactyls of almost equal length; sex organs at the first and second nodes of fertile whorls; no sex organ in the sterile whorls; antheridium light orange, 360 to 400 μm in diameter, oogonium brown, 400 to 430 μm in length and 350 to 360 μm in breadth, 8 spiral cell convolutions.

Collection No. 218; Holotype at author's laboratory, Isotype No. 50 at B.S.I., CBE, T.N.

4. *Nitella pseudoflabellata* var. *delicaluta* var. nov. D. Subramanian, 2005, National Seminar on Present Scenario in Plant Science Research, Abs. No. 1.2, pp. 7&8. (Plate - 1c).

A form of this species collected is dwarf, 4 to 6 cm in height, light green, monoecious, basal sterile whorls larger and upper fertile whorls 3 to 4 covered with mucilage clouds, smaller; sterile whorl into 6 primary rays, each into 7 secondary rays, each secondary ray into 7 tertiary rays; each tertiary ray into 1 or 2 quaternary rays and 3 or 4 dactyls, dactyl 2 celled, tip cell colourless and smaller; fertile whorls 2 times furcated; sex organs at the nodes of primary and secondary rays, sex organs rarely present in the nodes of primary and secondary rays of the sterile whorls also; antheridium orange, 300 to 320 μm in diameter; oogonium yellowish green; 320 to 340 μm in length and 280 to 300 μm in breadth; 8 spiral wall convolutions.

Collection No. 221. Holotype at author's laboratory; Isotype No. 71 at B.S.I., CBE, T.N.

5. ***Nitella tenuissima* (Desv.) Kutz. Em var. *puduchattiransis* var. nov. D. Subramanian, 2005. National Seminar on Present Scenario in Plant Science Research, Abs. No. 1.2, pp. 7&8. (Plate - 2c).**

Tiny plant to 3 to 3.5 cm high, monoecious, branchlet whorls forming beaded appearance, one above the other; at each node 6 primary rays forming 2/3 of total length of branchlets, each primary ray into 6 secondary rays of which the central two longer, peripheral secondary ray into 6 tertiary rays, central tertiary ray into 4 quaternary rays, each of which ending into 4 dactyls; peripheral tertiary ray into 1 or 2 quaternary rays and 2 to 3 dactyls; the quaternary ray ending into 2 to 3 dactyls; dactyls 2 celled; antheridia grey coloured, 200 to 220 μm in diameters; oogonium 220 to 240 μm in length and 180 to 200 μm in breadth, 8 spiral wall convolutions.

Collection No. 224, Holotype at author's laboratory and Isotype No. 51 at B.S.I., CBE, T.N. (Plate – 6).

6. ***Nitella furcata* subsp. *fruiticosa* subsp. nov. D. Subramanian, 2005, National Seminar on Present Scenario in plant Science Research, Abs. No. 1.2 : pp. 7&8. (Plate - 2a)**

Plants monoecious, to 10 to 12 cm high, light green and medium sized; two types of whorls, fertile and sterile; lower whorls sterile without sex organs; upper sterile whorls having rarely solitary sex organs, sterile whorls into 5 primary rays each into 6 secondary rays, each secondary ray into 2 to 3 dactyls and 1 or 2 tertiary rays ending into 2 to 3 dactyls, dactyl 2 celled; fertile whorls arise at the nodes of primary rays; the axillary branches into 1 or 2 sterile whorls and then into 1 or 2 fertile whorls; all the terminal 2 to 3 whorls smaller and fertile; all the fertile whorls enclosed in mucilage clouds; fertile whorls smaller or larger depending upon the number of furcations of fertile whorls; smaller fertile whorls 2 times furcated and larger fertile whorls 3 times furcated; sex organs at the nodes of primary and secondary rays; antheridium light orange, 300 to 350 μm in diameter, oogonium light brown, 320 to 360 μm in length and 280 to 300 μm in breadth; 8 spiral cell convolutions.

Collection No. 227; Holotype at author's laboratory and Isotype No. 52 at B.S.I., CBE, T.N.

II. MUSILAGE CLOUDS

Some of the species of *Nitella* have mucilage clouds on the uppermost whorls of the thallus, or axillary fertile whorls. But, the author has collected some *Nitella hyalina* (Plates 4 & 4a) plants after heavy floods due to heavy rainfalls, in his native place Ko. Athanur, near Virudhachalam Taluk of Cuddalore District 10 years back and they were completely enveloped with mucous clouds. It was very difficult to collect and handle them. Before the flood, that particular species had not been observed in the village. Therefore, the heavy flood should have brought them to my village from somewhere else. After that year, the author could not see those plants during the rainy reasons in the next years, eventhough *Chara* species alone (*Chara zeylanica* and *C. fibrosa*) are present permanently in the village.

After this, some 15 years ago, the author collected 2 varieties of *Nitella myriotrichia* (Plates 4b & 4c) in a paddy cultivation (irrigated) at Kummidipoondi, North to Madras. They were also slippery and highly viscous. The author could not catch them with hands. They were named as 1. var. *acuminata* and 2. var. *recurvata* and the author has published them in his book Monograph on Indian Charophyta in 2002.

Recently Mr. Sheshadri Ramaswamy, a nature lover from Bangalore has sent the author 2 liquid preserved specimens from Mangalore. They were *N. hyalina* (Plate 4d) and *N. pseudofl abellata*. Both of them were completely made up of mucilage cell walls. Hence it was not possible to catch them with hands. They were highly viscous and delicate.

So far, the author has no informations about the entirely viscous Charophyta. The partly or wholly enveloped plants with mucous clouds are an adaptation to protect themselves from heavy water current or heavy temperature or other external disturbances.

III. ROBUST AND SPINY FORMS

The next defencive mechanism is the presence of robust form of the thallus as in some forms of *Chara* due to over development of spine stipules, stipulodes, bract and bracteols, like *C. canescens f. hirsuta* T.F.A. and *C.hirsuta* T.F.A. The other defencing mechanism is the spiny forms of thallus as in *Chara hornemannii* Wallm. em. (Plate 5) and *C. hornemannii f. nordhoffiae* (T.F.A.) R.D.W. (*C. hornemannii nordhoffiae* T.F.A.) (Plate 6). Pundhir, H.S. (1994) has collected from North India a form of this species namely *Chara hornemannii var. longifolia* (Roxb) R.D.W. and studied the cytological characters. *Chara zeylanica* a robust form was collected near Sriperumpudur near Madras and the same species in salt water ditches near Mandapam and Pamban bridge, Kille and Pichavaram collected by the author are other examples for robust forms of Charophytes (Plate 7).

IV. TERRESTRIAL NITELLA

The next ecological adaptation of Charophyta is the occurrence of terrestrial *Nitella*. M.O.P. Iyengar, the world famous Algologist of India has collected some tiny balls like green bodies on moist grounds of Kodambakkam, Mahabalipuram, Vandalore, etc. from 1930 to 1950. He tried to make a correct identification of the Algae with famous British Algologist Prof. J. Groves but in vein. Prof. Iyengar named it as *Nitella terriestris*. The basal node of stem repeatedly branched to give a dwarf bushy appearance, some 1 to 1.2 cm in diameter. The sterile and fertile whorls are separate. The antheridia are quadriscutate. When these plants were partly submerged under water, as observed near a small lake of Vandalore, the plants are larger and taller. When these plants were totally submerged under water, they became the tallest and largest plants. Therefore, terrestrial forms of this species adapted themselves into a smaller, more condensed green ball like forms. This is to withstand hot climatic conditions and unusual habitat in which they are living. The light intensity is more in soil than inside water. To obtain more light the submerged water plants may assume larger size with more relaxed branches. This *N. terrestris* (Plates 8 & 9) is the first terrestrial plant of the world. The author has seen it in a bottle with 4%

liquid formalin preservation some 15 years back. He is trying to find it not only from Madras but also in other parts of Tamil Nadu.

Likewise, the other Algologists may try to find out *N. Terrestris*found by M.O.P. Iyengar from Tamil Nadu.

Finally, some 15 years back there was a heavy flood in Tamil Nadu. The author was garden incharge of Botanical Garden, AnnamalaiUniversity. After the flood was over, the author could find some green plants spreading on the moist ground under shade of trees. The soil was cleyish sandy. The author carefully collected them and identified as *Nitella* species. The plants were rooting at the nodes upto the middle of the plant. The plants were 4 to 5 cm long and when removed from the soil, they broke into small portions. The author reconstructed them and described them in his book as *N. Kangambigai* D.S. and *N. Kollachiappansis* D.S. and published in a journal Phykos (1997) and finally in his book "Monograph on Indian Charophytes". These plants after flood, spreaded on ground due to slow motion of water and became spread on the soil and roots were produced in each and every node upto almost ½ of the portion of the plant. These species also have quadriscutate antherida like that of *N. terrestris*.

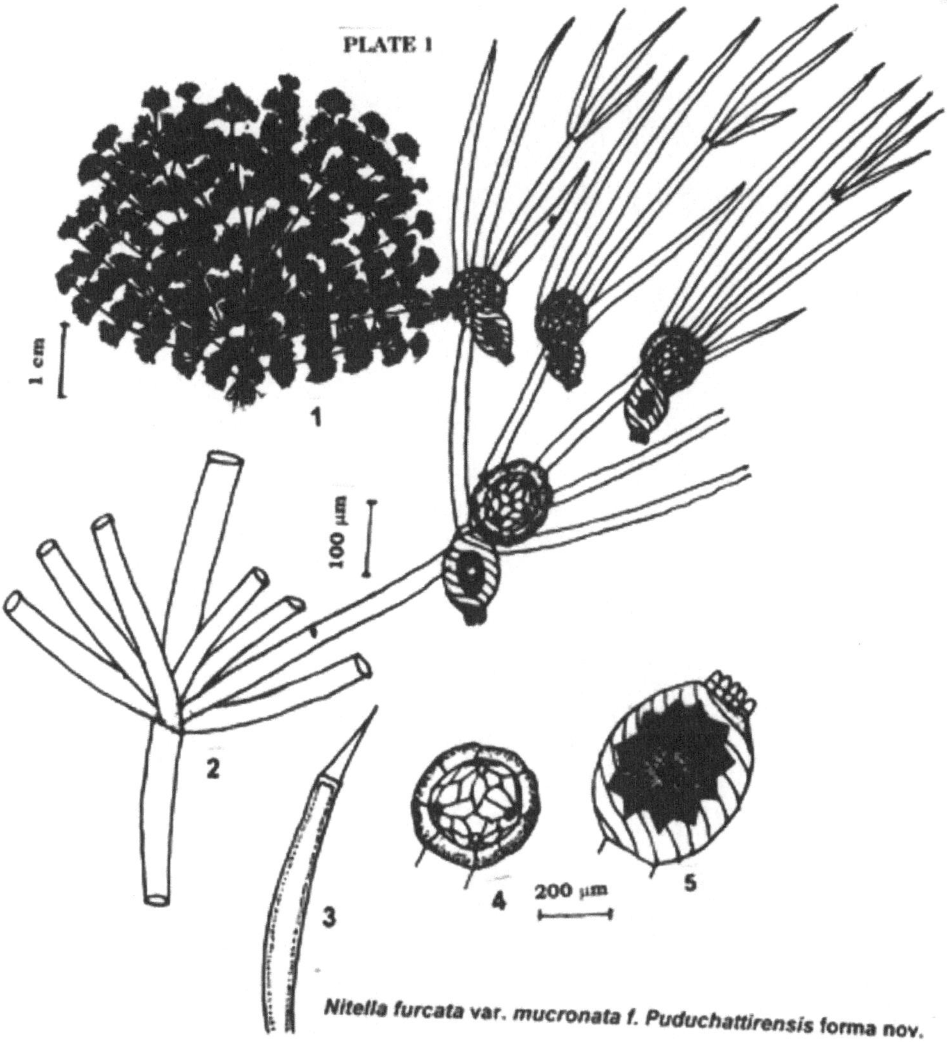

Plate 1 *Nitella furcata var. mucronata f. Puduchattirensis* forma nov Fig 1 Habit; 2= A whorl enlarged; 3=A dactyl; 4 = Antheridium; 5 = Oogonium.

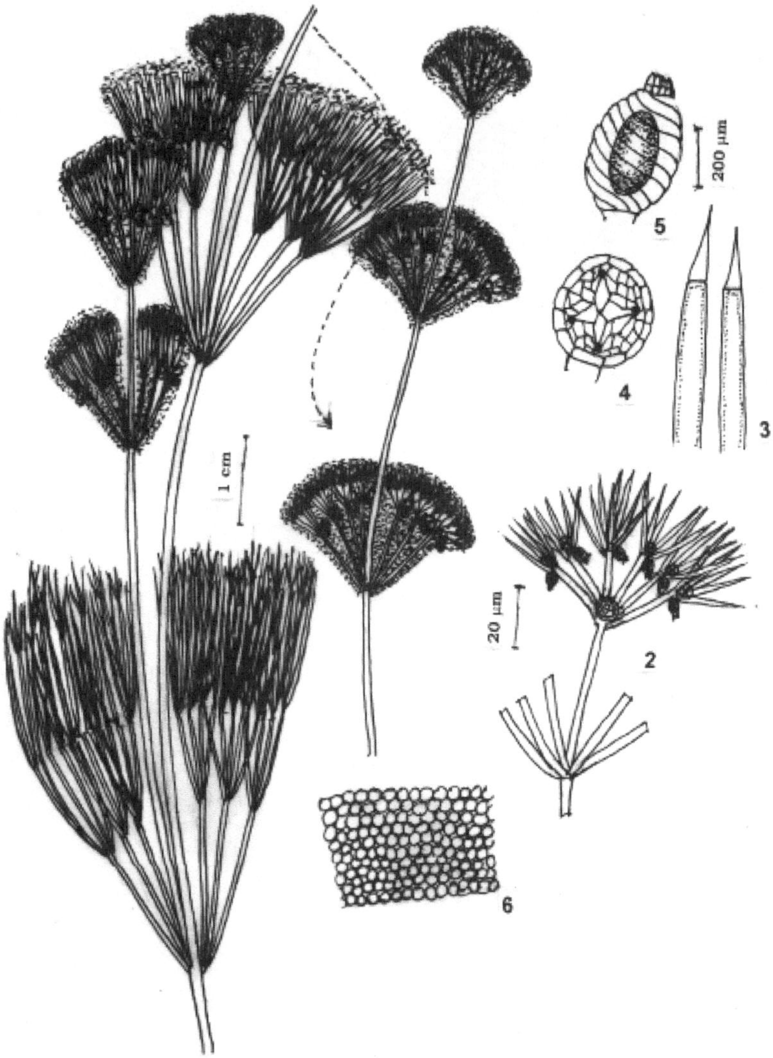

Plate 1a *N.pseudoflabellata* var. *angelli* var.nov. Fig.1 = Habit enlarged; 2 = A whorl enlarged; 3 = Dactyls; 4 = Antheridium; 5 = Oogonium.

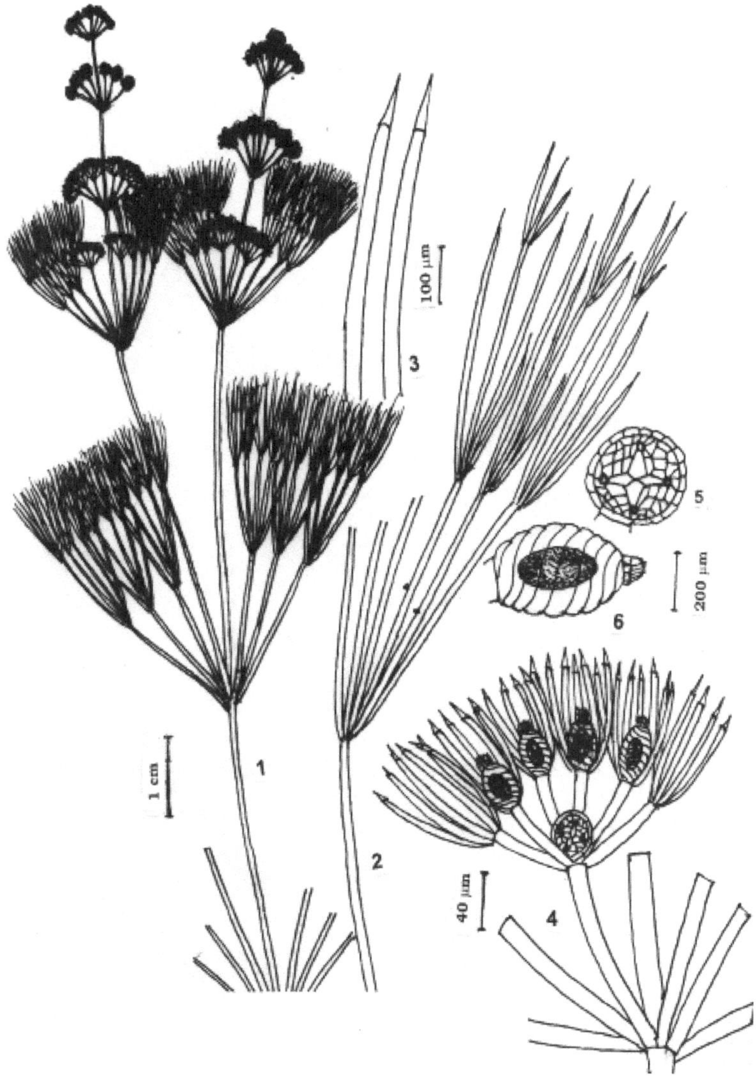

Plate 1b *N.pseudoflabellata* **var. *pantherii* var. Nov.** Fig.1 = Habit; 2 = A sterlie whorl; 3 = Dactyls; 4 = A fertile whorl; 5 = Antheridium; 6 = Oogonium

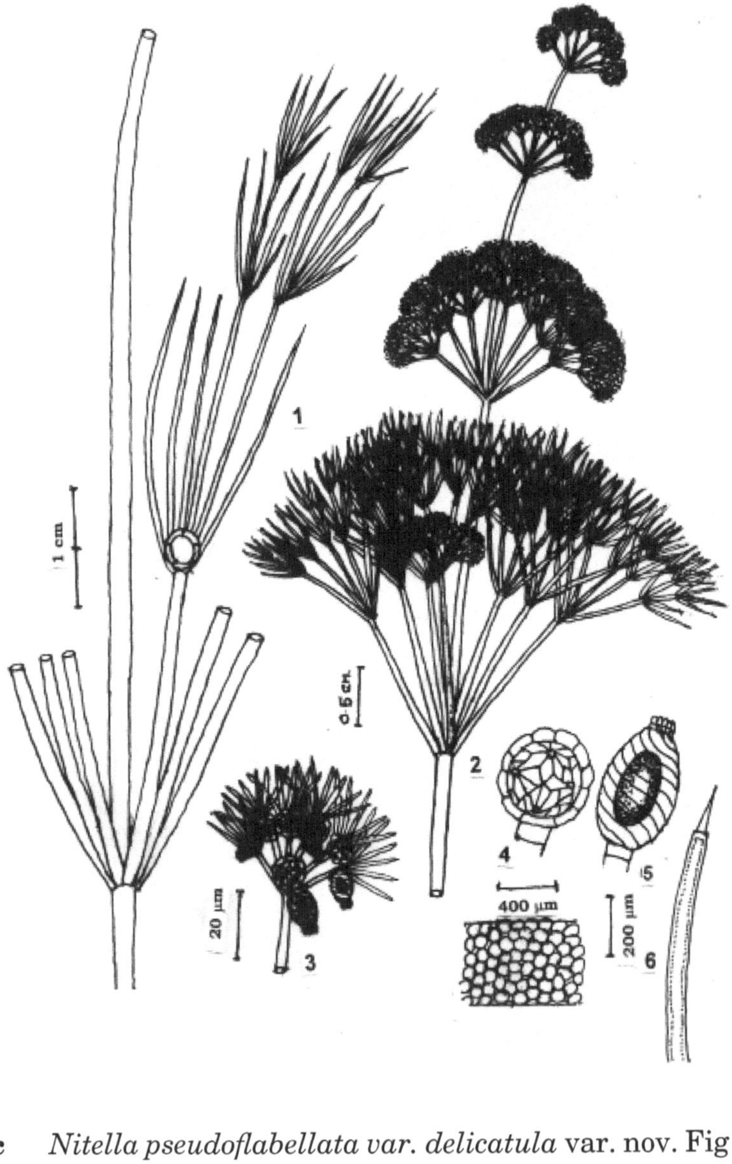

Plate 1c *Nitella pseudoflabellata var. delicatula* var. nov. Fig 1 = Basal whorl; 2 = Distal whorls; 3 = fertile whorl enlarged; 4 = Antheridium; 5 = Oogonium; 6 = Dactyl; 7 = Oospore membrane

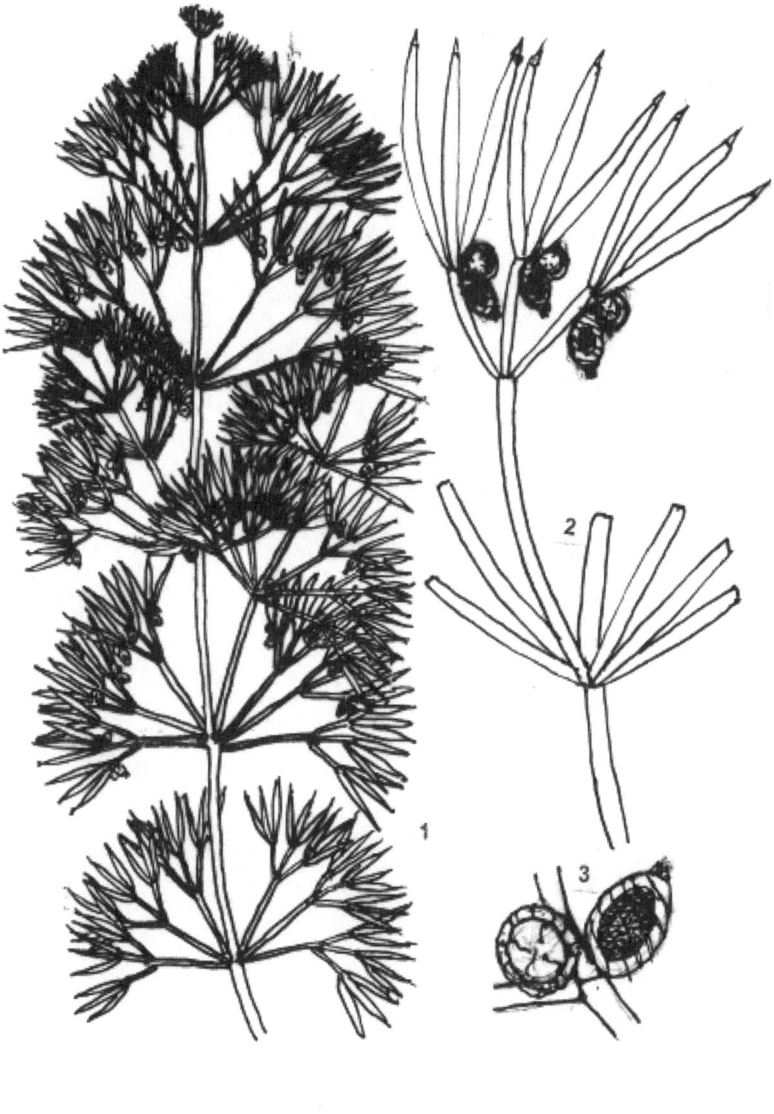

Plate 2 *Nitella furcata* – Normal Plant

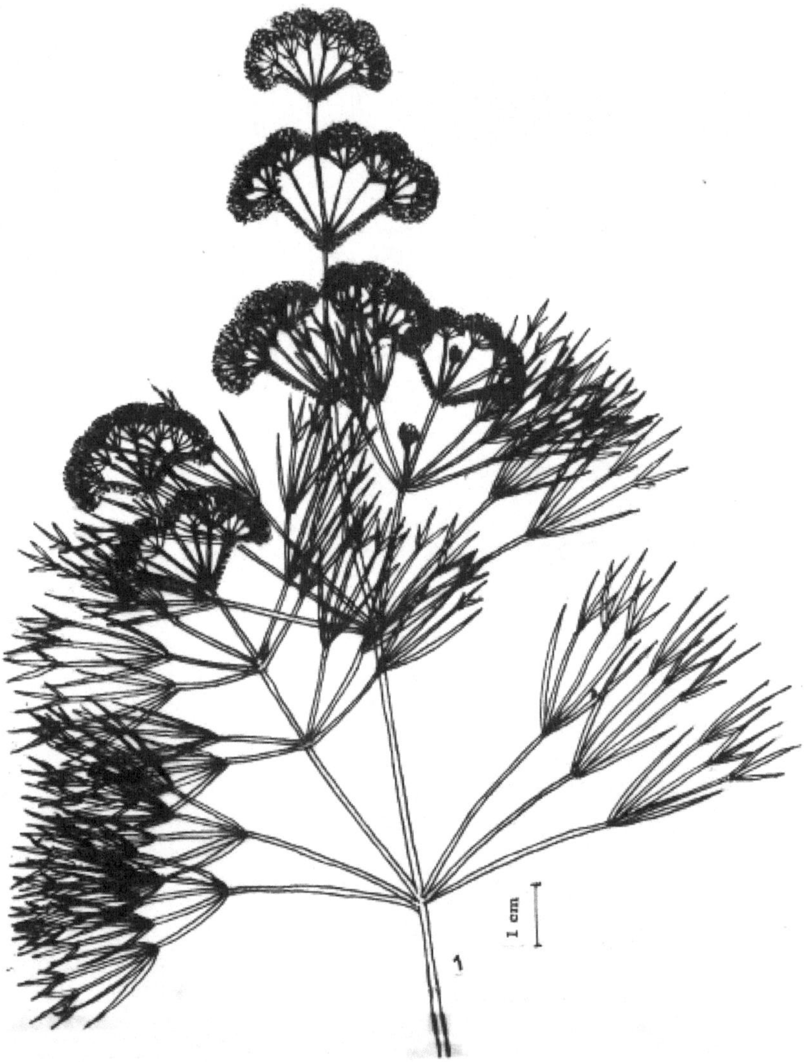

Plate 2a *Nitella furcata* sub sp. *fruiticosa* subsp.Nov.

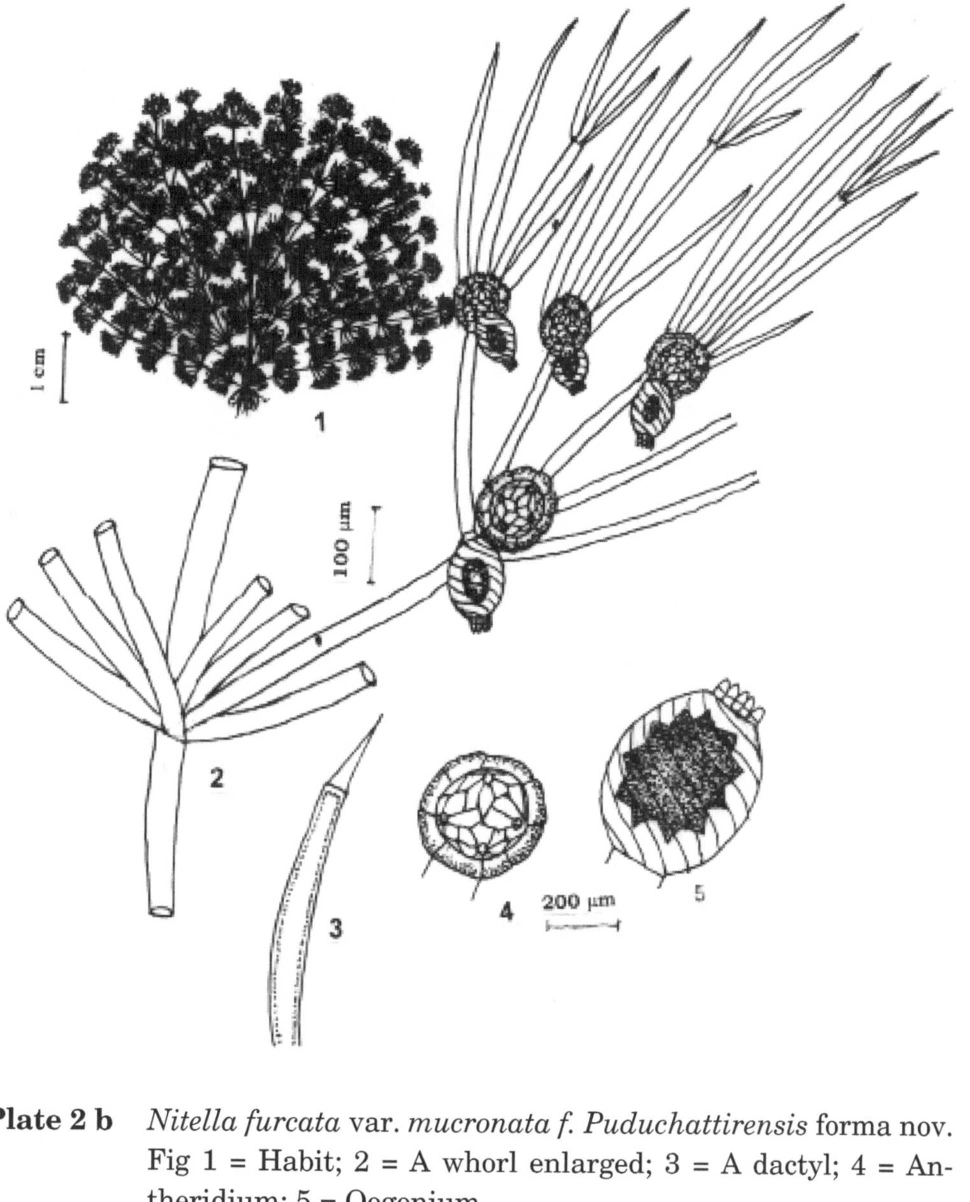

Plate 2 b *Nitella furcata* var. *mucronata f. Puduchattirensis* forma nov. Fig 1 = Habit; 2 = A whorl enlarged; 3 = A dactyl; 4 = Antheridium; 5 = Oogonium.

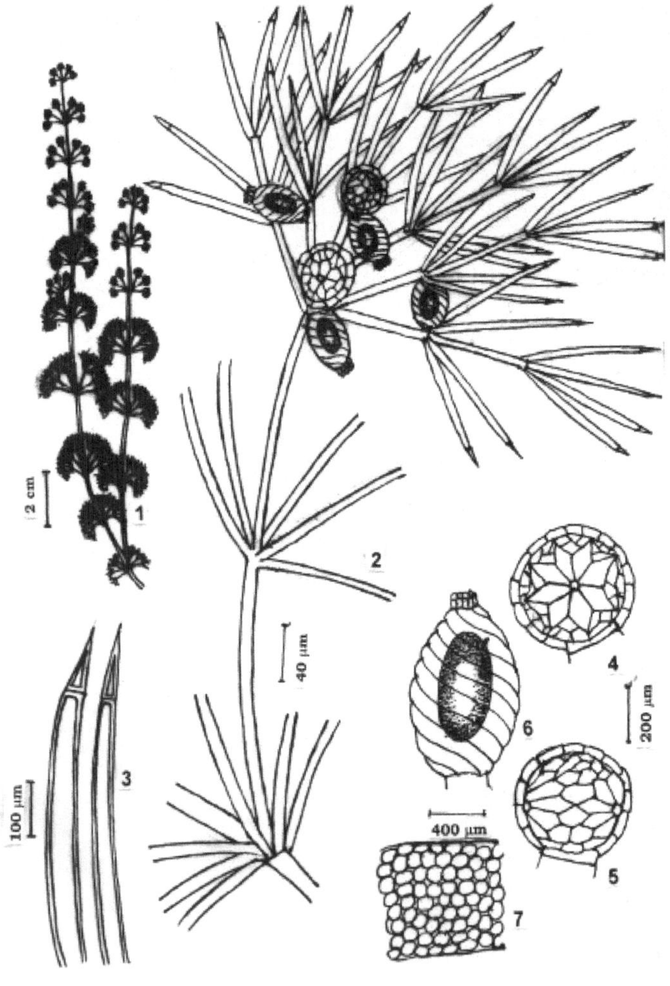

Plate 2 c *Nitella tenuissima* var. *Puduchattirensis* var. nov. Fig 1 = A plant; 2 = whorl enlarged; 3 = dactyls; 4 and 5 = Antheridium; 6 = Oogonium; 7 = Oospore membrane.

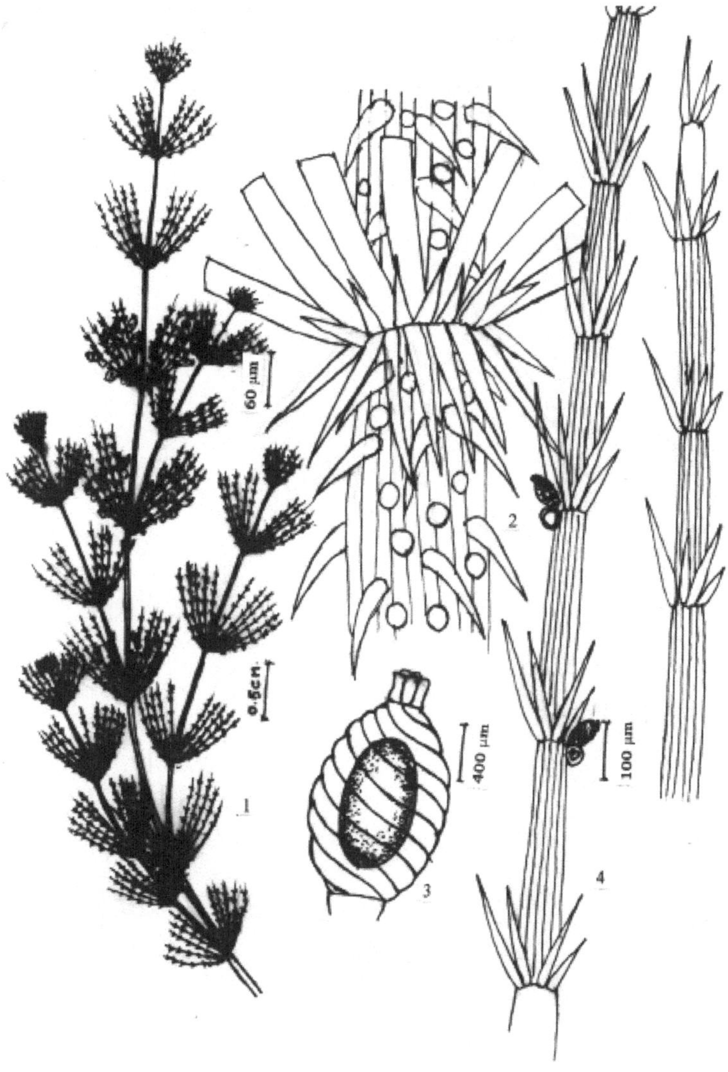

Plate 3 *Chara fibrosa* – Normal Plant Fig 1 = A plant; 2 = A node with stipules and stipulodes; 3 = Oogonium; 4 & 5 branch of limited growth

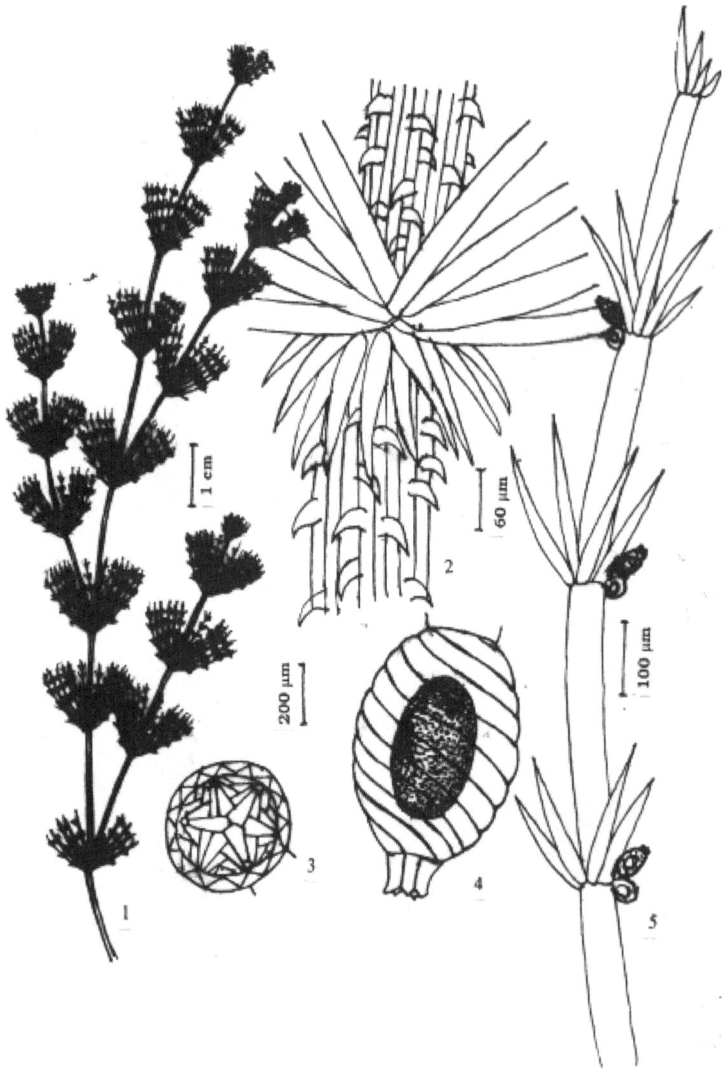

Plate 3a *Chara fibrosa* – Modified Plant Fig 1 = A plant; 2 = A node; 3 = Antheridium; 4 = Oogonium; 5 = A branch of limited growth.

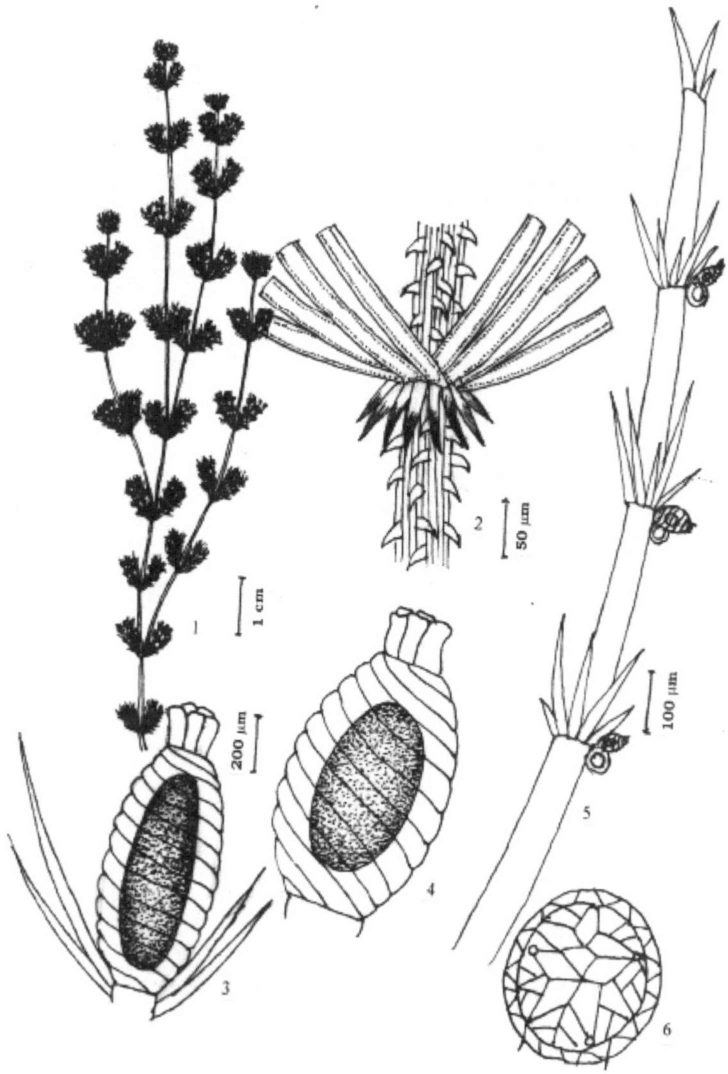

Plate 3b *Chara fibrosa* – Modified Plant Fig 1 = A plant; 2 = A node; 3 = Young oogonium; 4 = mature oogonium; 5 = A branchlet; 6 = Antheridium.

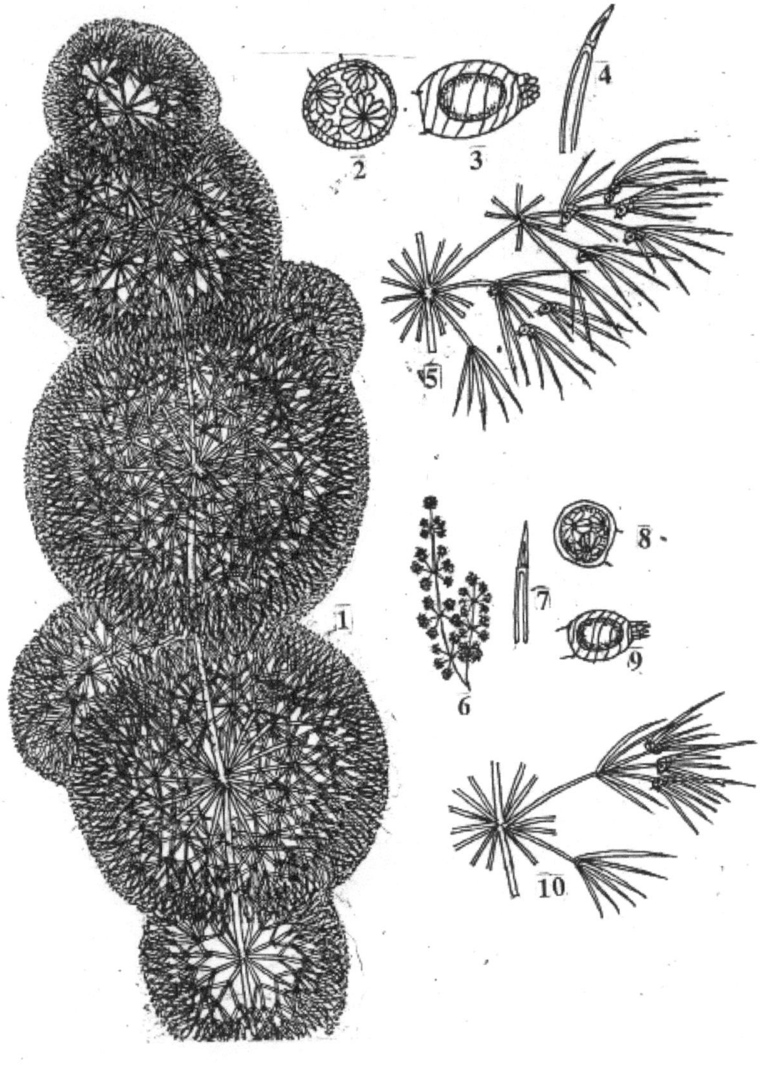

Plate 4 *Nitella hyalina* – var. *athanarensis* Fig 1 = Plant; 2 = Antheridium, 3 = Oogonium, 4 = Dactyl; 5 = 3 types of whorls Figs 6 to 10 = *Nitella lhotzkeyi* – reduced from Fig 6 = Plant; 7 = Dactyl; 8 = Antheridium; 9 = Oogonium; 10 – 2 types of whorls

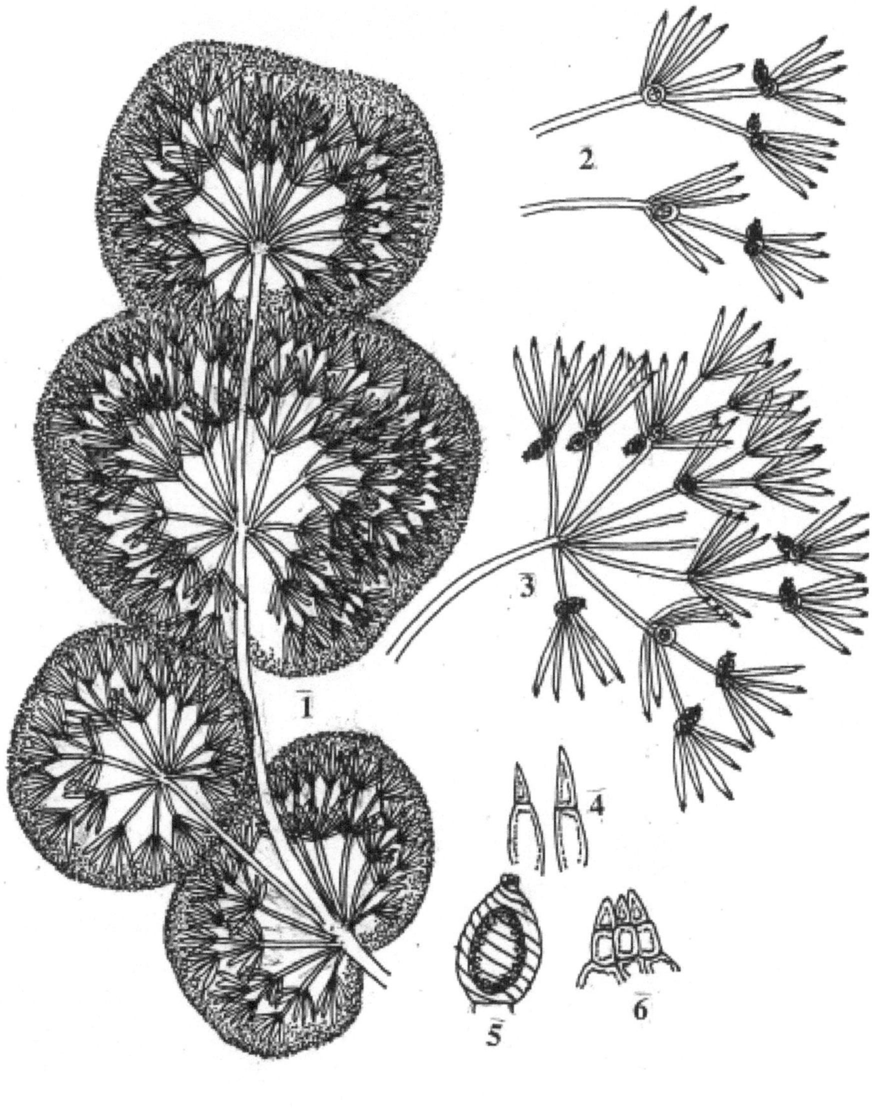

Plate 4a *Nitella hyalina* var. *chelliammai* – inside mucilage clouds

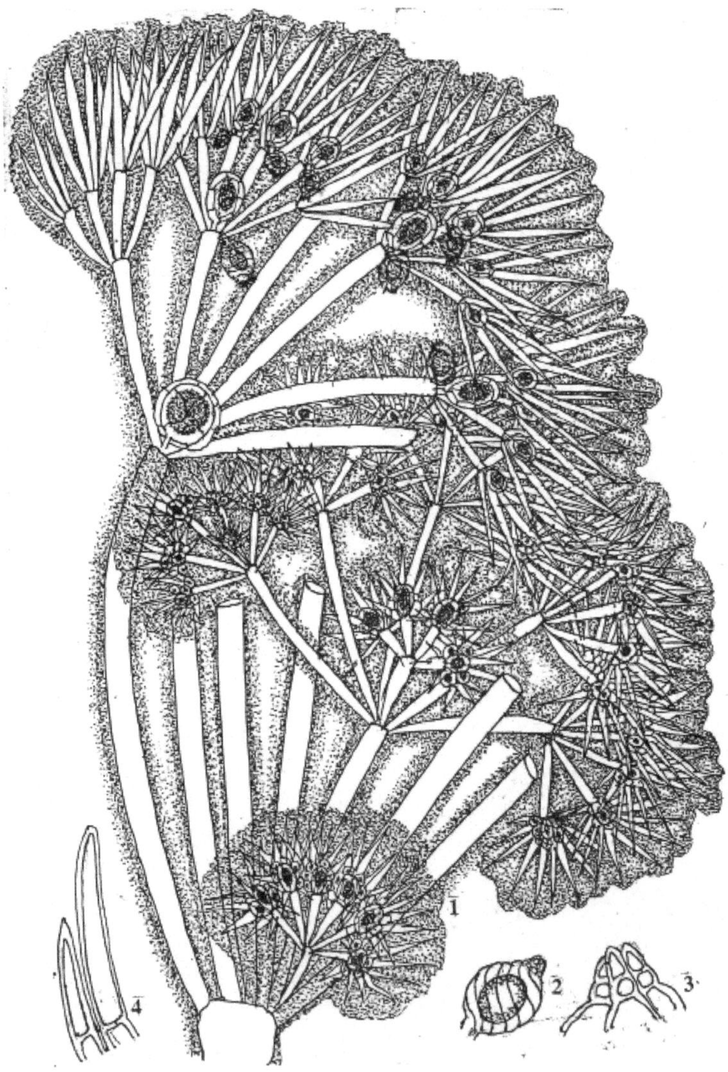

Plate 4b *Nitella myriotrichia* var. *acuminata* 1 = part of plant; 2 = Oogonium; 3 = coronula cells; Dactyl

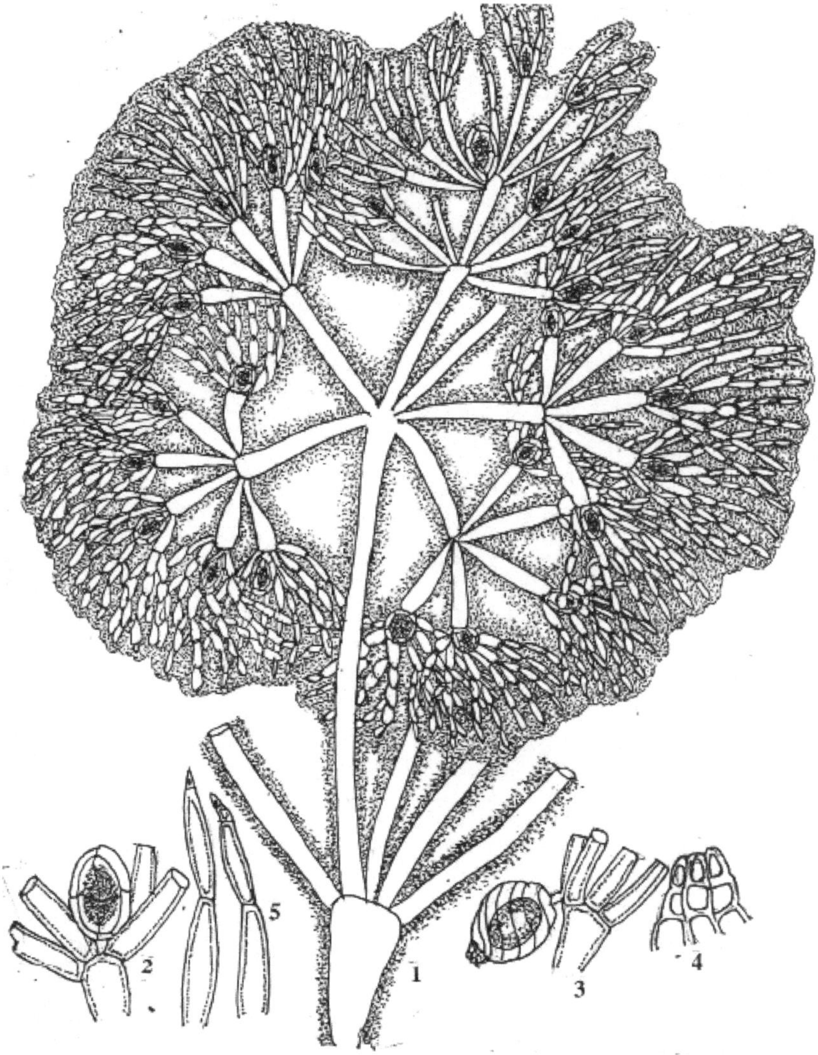

Plate 4c *Nitella myriotrichia* var. *incurvata* 1 = Part of plant; 2 = Antheridium; 3 = Oogonium; 4 = Coronula cells; 5 = Dactyls

Plate 4d Figs 1 to 8 = *Nitella hyalina* viscous plant Fig 1 = A plant; 2 = A portion enlarged, 3 to 5 = 3 types of whorls; 6 & 7 = Antheridia; 8 = Oogonium Fig 9 to 10 = Antheridia and Oogonium of *Nitella pseudoflabellata*

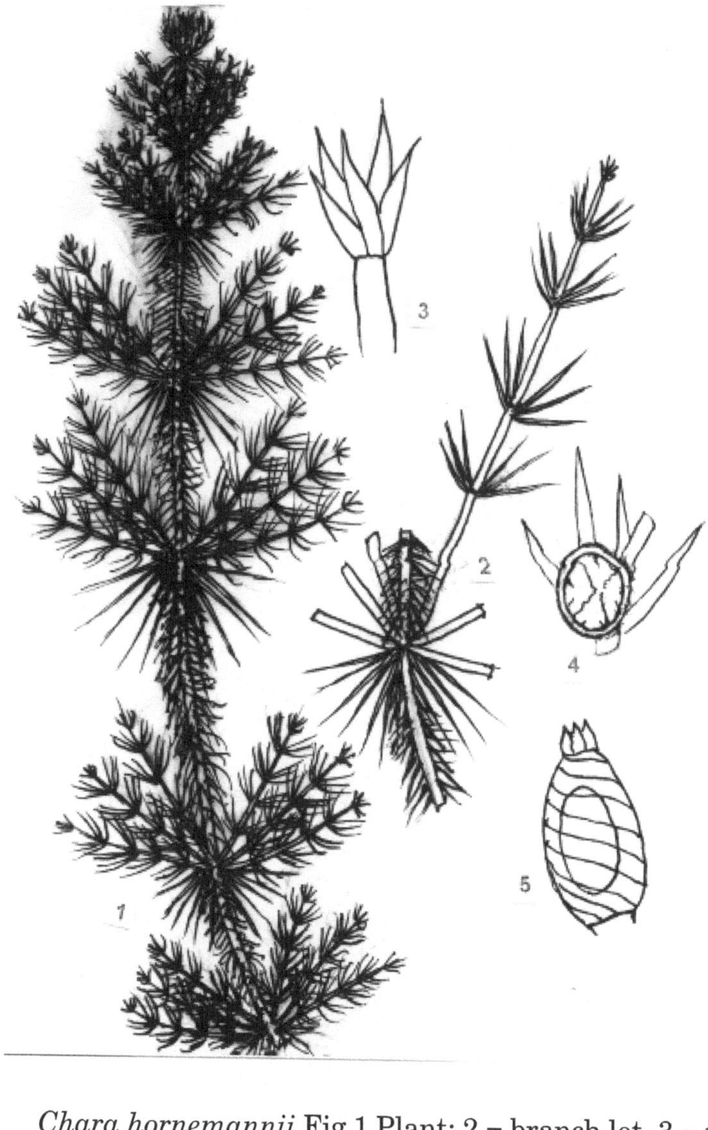

Plate 5 *Chara hornemannii* Fig 1 Plant; 2 = branch let, 3 = tip cells; 4 = Antheridium; 5 = Oogonium

Plate 6 *Chara hormemannii f. norhoffiae*

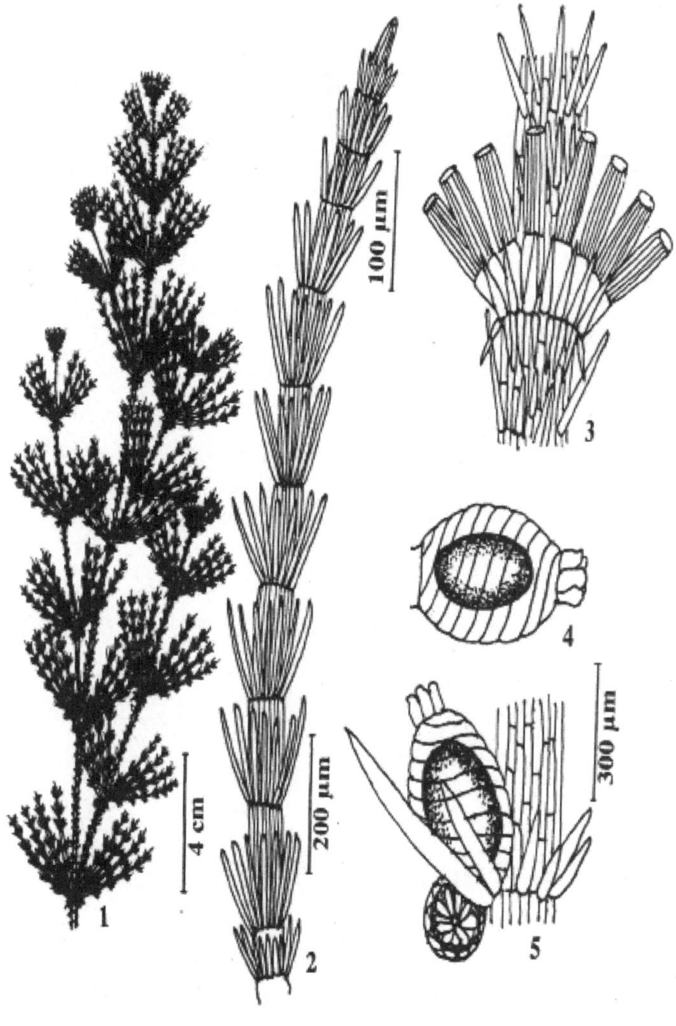

Plate 7 Figs 1-5: *Chara zeylanica* (Salt water form). 1. Morphology of the plant; 2. A branchlet; 3. Axis node; 4. Oogonium; 5. A branchlet node with sex organs.

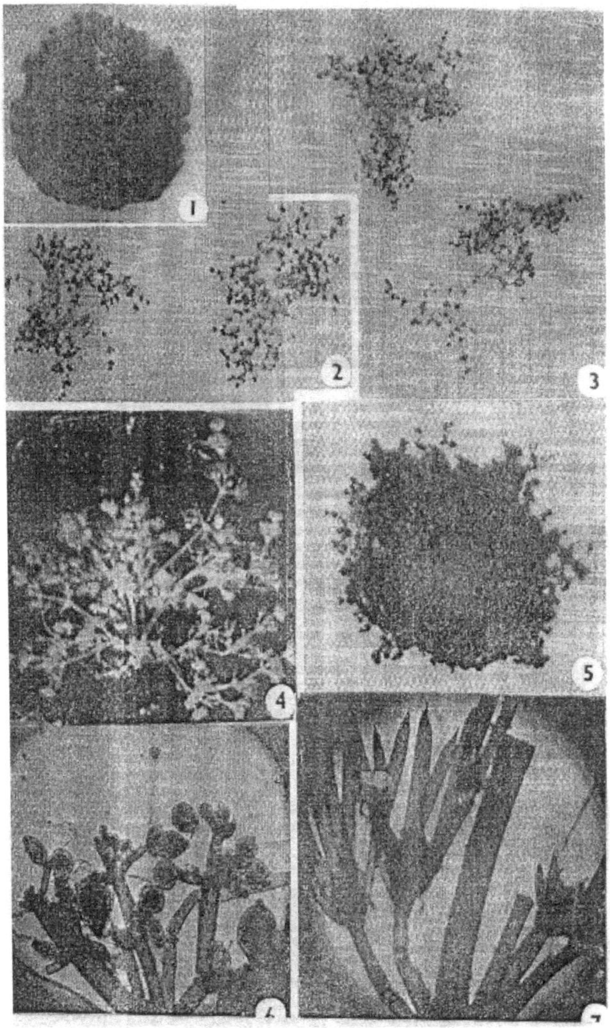

Plate 8 *Nitella terestris. Photos* Fig 1 = A plant in a dense cluster on moist ground; 2 and 3 = portions of the plant showing in Fig. 5, teased and spread out to show its habits; 4 = portion of a single plant with a number of primary axis arising from a common node; 5 = An older plant growing slightly submerged, showing its more or less lax and spread out habit; 6 = Axillary fertile branch showing gametangia; 7 = stem node showing sterile and fertile branchlets; Fig 1 = 3; Figs 2, 3, 5 = 2; Fig 4 =3; Figs 6 = 7 & = x35.

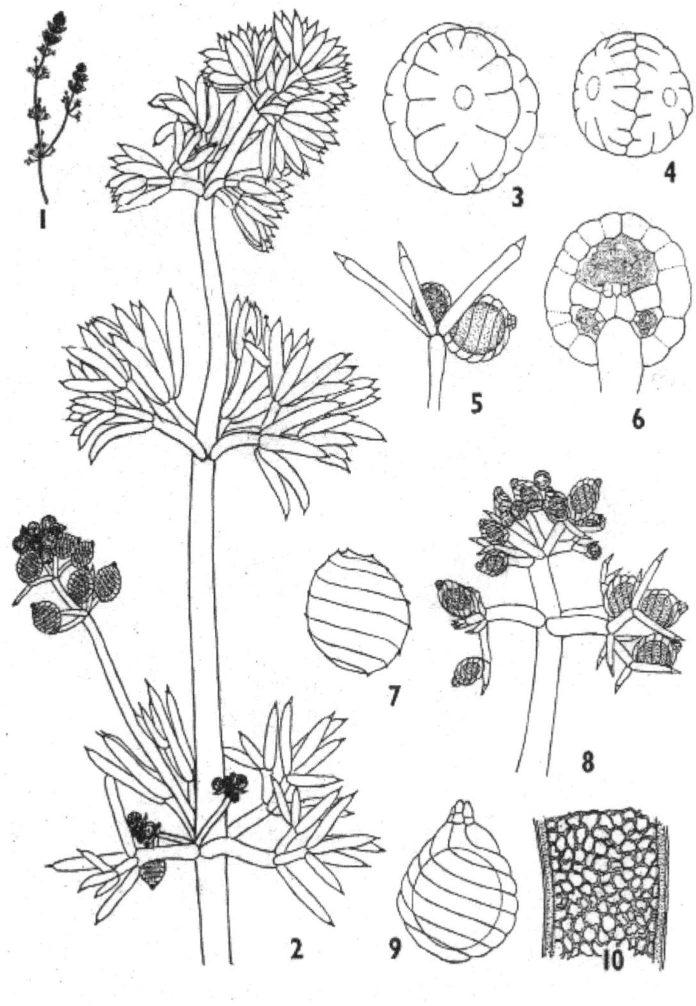

Plate 9 *Nitella terestris* Diagrams Fig 1 = A small branch of a plant growing in trickling water x 2 ; 2 = A portion of a primary axis, showing sterile and fertile whorls x 40; 3, 4,6 = Antheridia; 3, 4 = surface view of antheridia showing the shields in two different views x 730; 5 = Gametangia x 195; 6 Median optical section of an antheridium x 730; 7 = Oospore x 200; 8 = A fertile axillary branch showing gametangia x 100; 9 = An oogonium x 200; 10 = Reticulate markings on the Oospore membrane x 1500

2. CHAROPHYTES FROM TAMILNADU

1. ***Nitella congesta A.Br. Var filiformis*** var. nov. D. Subramanian, 2009 in Diversity of Charophytes in South India. National Seminar on Climate Change and Environmental Challenges. Article 2.13, p. 33 (Plate - 10).

It is a peculiar species of *Nitella* collected at Kanyakumari District of Tamilnadu.

Plants thin and filiform, longer, 20 to 25 cm tall, rarely branched, internodes many times (4 to 5 times) longer than branchlet whorls. It is dioecious, male and female plants separate, only female plants alone seen; 5 branchlets at each node of axis and branches. Each primary ray forks into 3 to 4 dactyls and 1 secondary ray, which again forking into 2 to 3 dactyls. Sex organs at the nodes of primary and secondary rays, dactyls 2 celled, end cell short and acute, very rarely sterile branchlet may be also present. This is a very simple species of *Nitella* studied by the author so far.

This species resembles to some extent to *N. congesta* (R.Br.). A.Br. described by Wood and Imahori (1964). But in this species studied by the author, all the fertile branchlets are similar and sterile whorls very rare. Fertile whorls simple and not so much crowded when compared to *N. congesta*. The general morphology of the present plant is totally different from *N. congesta*. It is also resembling to *N. substilissima* but differs from itdue to the large size of plant and nature of whorls. At present it is treated as a variety of *N. congesta* and named here in *N. congesta* var. *filiformis var*. nov.

Collection No. 235. Holotype at author's laboratory and Isotype, No. 72 at B.S.I., CBE, T.N.

2. *Nitella furcata var. grandiformis*. D. Subramanian in Diversity of Charophytes in South India. National Seminar on Climate Change and Environmental Challenges Article 2.13, p. 33. (Plate - 11).

This plant was collected in a small ditch by the side of small hills, near Madras. Holotype 237. Isotype No. 54 at HerbariumCenter B.S.I. Coimbatore.

Plants slender, small and delicate, rarely branched; 6 to 8 cm tall, internodes of axis longer than branchlet whorls. At each node of axis, 7 branchlets arise, all equal and only one type; Each branchlet forks into 6 to 7 primary rays, which are 3/4 of the entire length of the whorl; each primary ray into 4 to 5 secondary rays, each of which ending into 3 to 4 dactyls. No sex organs observed.

In general appearance, this species is mistaken as *Nitella hyalina*. But, when examined under the microscope, unlike the 3 types of whorls in *N. hyalina* in each node of main axis, or two types of whorls as in *N. lhotskey*, this species has all the whorls are of only one type and equal in size. The fureations of branchlets resemble to *N. furcata* and so it may be a variety of *N. furcata* resembling *N. hyalina* and *N. lhotskeyi* and named here in as *N. furcata* var. grandiformis var. nov.

But, the plants are as beautiful as *N. hyalina*. Perhaps, from *N. hyalina, N. lhotskeyi,* should have originated and from such a plant like *N. lhotskeyi*, this plant, *N. furcata var. grandiformis*var. nov. should have been evolved by reduction of second and third types of whorls in *N. hyalina* and second type of whorl in *N. lhotskeyi*. All these three members are related together.

3. *Nitella lhotskeyi* A.Br. *f. hyalinoides* D. Subramanian var. nov. (Plate - 12).

The plants were collected at Thirunelveli of Tamilnadu in a small natural ditch.

Plants 5 to 6 cm tall rarely branched, almost erect, only 2 types of branchlets present, both with sex organs.

Large whorls, each with 8 primary rays, 2/3 long of the entire branchlet length. Each primary ray into 2 dactyls and 3 secondary rays, each of which into 3 to 4 dactyls of equal length, small whorls with 2 dactyls and 3 secondary rays, each of which ending into 3 to 4 dactyls.

Wood and Imahori (1964) have described 2 types of plants in this species, the type species and a form of this species *N. lhotskeyi f. substilissimoides* R.D.W. In general morphology and details of fertile and sterile whorls, the present plant is almost distinct when compared to the above 2 types of plants.

As the present plant studied by the author resembles *N. hyalina* in general appearance, it is treated as a form, namely *N. lhotskeyi. f. hyalinoides,* D. Subramanian.

Collection No. 233, Holotype and voucher specimens are deposited at 277, Mariappa Nagar, Annamalai Nagar – 608 002, Tamilnadu; Isotype No. 53 at B.S.I., CBE, T.N.

4. *N. lhotskeyi f. minima forma* nov. D. Subramanian (Plate -13).

This plant was mistakenly described as *N. hyalina* var. *puduchattirensis* D. Subramanian in the author's previous book "Monograph on Indian Charophyta."

When re-examined by the author it turned out to be another species, *N. lhotskeyi* and then as this plant was too small to recognize easily in the middle of other Charophytes and Angiosperm weeds and as the fertile and sterile branchlets are very simple, it is named here in as *N. lhotskeyi. f. minima.* var. nov.

Plants very small, 2 to 3 cm tall, monoecious, only tip of the plant enveloped with mucus clouds. Stem 220 to 230 μ thick, whorls of 6 fertile and large and 6 sterile and small branchlets; each of the fertile branchlets forks into 2 dactyls and 3 secondary rays and each of the secondary rays forks into 3 to 4 dactyls. Each of the sterile branchlets forks into 6 dactyls; dactyl

2 celled, internode shorter than branchlets. Primary rays 2/3 of the entire length of branchlets, basal cell of dactyl longer and slightly curved; Sex organs only in the second furcation of fertile whorls.

Antheridium orange and 500 μ in diameter. Oogonium orange brown 435 μ in length and 330 μ in breadth. Corona cells very small, 10 convolutions, oospore membrane finely granulate.

Collection No. 240,Holotype and voucher specimens are deposited at author's laboratory at Mariappa Nagar, Annamalai Nagar – 608 002, Tamilnadu. Collected in sandy soil with 1 foot water level in an Agriculture land near Puduchathiram, Cuddalore District by the author and Isotype No. 73 at B.S.I., CBE, T.N.

5. *N. hyalina* (DC) Ag. (Plates 14 and 15)

There are only two members described by Wood and Imahori namely 1) *N. hyalina* (DC) Ag. (Icone 349) and 2)*N. hyalina f. formosa* (T.F.A.) R.D.W. (Icone 350). It shows that this species is very rare in this world. But, the author has studied several types of this species in Tamilnadu, India. The author has described in his book in 2002 "Monograph on Indian Charophyta 3 district plants 1) *N. hyalina* var. *chelliammai* D. Subramanian, 2) *N. hyalina* var. *athanurensis* D. Subramanian and *N. hyalina* (DC). Ag. Mavidanthal form. *N. hyalina* var. *puduchattirensis* has been now changed into *N. lhotskeyi* var. *minima* var. nov. Further 3 forms of *N. hyalina* have been described in this book from Thiruvannamalai of Tamilnadu. Apart from these there are a large number of distinct forms of this species collected from various places of Tamilnadu, like Managudi of Thirunelveli District, Tuticorin, V.O.C. District, Thenkasi, Thirunelveli District, Trivellore, Chengai. M.G.R. District and Chittoor District of Andhra Pradesh, Kollidam of Tanjore District and Sankarapuram of Villupuram District of Tamilnadu. This shows that we may get plenty of materials of this species, if we make intensive field trips to various places of Tamilnadu, particularly hill resorts and red soil areas. Of all the species of *Nitella*, *N. hyalina*, *N. lhotskeyi*, *N. congesta* A.Br. and*N. substilissima* are very beautiful plants under the microscope at low power 40x or 20x of Dissection microscope. All these forms may be distinct from the type species, as they are living in different soil and environmental conditions.

6. *Chara corallina* Klein ex. Willd, em. (Plate - 16)

Plants collected from neighbouring areas of V.K.P. District, Tamil Nadu. Herbarium specimens are deposited at author's laboratory at 277, Mariappa Nagar, Annamalai Nagar – 608 002, Tamil Nadu.

Plants larger, 20 to 24 cm high, lower part of the plant unbranched, upper part more branched and crowded, giving a beautiful appearance in natural conditions, 6 branchlets at each node, main axis and branchlets all ecorticated, each branchlet 6 segmented, all spreading out, tip cell of branchlet 1, 2 or 4 celled, cells conical and pointed. No sex organs.

There are 8 members of *Chara corallina* described by Wood and Imahori with illustrations. Of these 8 plants, 4 are *C. braunii* like and 4 are *Protochara* like. In India and so many other countries of the world, only *C. braunii* like plants of *C. corallina* occur in large numbers. But, *Protochara* like plants of *C. corallina* occur only in New Zealand and Australia. But now, Wood and Imahori (1965) are not accepting *Protochara* as a genus. But, the author is of opinion that *Protochara* may be re-established with 2 species *P. australis* and *P. inflata*. These two species are not comparable to the *C. braunii* like species of *C. corallina* and these two species are also geographically isolated in a remote corner of the world, New Zealand and Australia, whether they are going to disappear from earth or re-establish and spread to other parts of the world is a question for which it is not easy to get an answer. Perhaps, detailed cytogenetical studies in the two groups of *C. corallina* will yield a fruitful result.

Another interesting point is that the large number of forms and varieties of this species in large quantities in each have been studied by the author in Tamil Nadu (like the present plant and *Chara corallina.* var.*portonovensis*D. Subramanian in "Monograph on Indian Charophyta" by the author in 2002)when compared to the 4 forms of*C. braunii*and *C. corallina* with illustrations by Wood and Imahori (1965).It shows that India in general and South India in particular are more suitable places with required conditions for the growth and development of this species.

7. *C. Corallina* Klein ex. Willd, em (Plate - 16)

The author collected on 17.01.2008 at Sathamangalam near Virudhachalam of Tamilnadu, 2 forms of *C. corallina* in a large fresh water ditch.

Form - 1

This is a large plant, 3/4 to 1 foot fall, highly branched and a large colony of nearly 100 plants may be found to a width of 2 to 3 feet.

Plants ecorticate, at each node of axis 6 branchlets, one axillary shoot and cluster of oogonia above and antheridia below formed; branchlet 5 segmented, at each basal node of branchlet, one oogonium above and one antheridium below present, basal segment smaller, no stipulodes, bracts and bracteoles smaller; Oogonia large, 850 to 860 μm long and 650 to 670 μm broad, spiral cells with 9 convolutions. Oospore brown and outer jacket orange; Antheridia 800 to 850 μm in diameter, orange.

Form - 2

This is resembling to the form – 1 described above in all respects excepting in smaller size of plants (8 to 12 cm tall), lesser compact and sex organs present only at the nodes of the main axis of the plant.

In general appearance the present 2 forms of *C. corallina* described above resemble *Chara zeylanica* var. *f. brittonii* (T.F.A. ex. Rob.) R.D.W. (*C. brittonii* T.F.A. ex. Rob.) described by Wood and Imahori in their book. But *C. zeylanica* var. *brittonii* is very small, 2 cm high, greyish green, tufted rather heavily incrusted, resembling a wave swept sand beach form of *C. vulgaris* or *C. globularis* and the present forms of *C. corallina* are larger, light green forming a colony of 40 to 50 plants to a width of 2 to 3 feet. In *C. zeylanica* var. *brittonii*, there are stipulodes in 2 tiers but in the present forms they are absent. The present forms resemble *C. corallina* var. *wallichii* (A.Br.) R.D.W. in general appearance but in all dimensions, the present forms are larger than *C. corallina* var. *wallichii*.

The soil is a mixture of calciferous and red soil and the water is pure and clean favouring the growth of a large number of *Chara* species like *C. corallina, C. braunii* and *C. fibrosa* and *Nitella flexilis, N. translucens, N. gracilis, N. furcata, N. hyalina* and *N. pseudoflabellata.*

There are a large number of different forms and varieties among the Indian species of Charophyta. In an abstract "Diversity of Charophytes in South India" a large number of forms, varieties and species of *Chara* and *Nitella* have been presented by the author in National Seminar on climate change and environmental challenges" held on 7 and 8, March 2009 at Botany Department, AnnamalaiUniversity, Tamilnadu. Some of the varieties and species of *Chara* and *Nitella* have been also published in journals like Glimpses of Indian Phycology, 2005, edited by Keshri, J.P. and Kargupta, A.N. and published by Bishen Singh Mahendra Pal Sing, Dehradun, India, Indian Hydrobiology in 2000 and Phykos, 2000.

Some of the members were collected by the author from various places of South India and the herbarium specimens have been deposited in the Author's laboratory at Mariappa Nagar, Annamalai Nagar – 608 002, Tamilnadu.

By observing under the low power and high power lenses of Compound Research Microscope,important characters of the plants have been drawn and the taxa were identified with the help of Monograph on "A Revision of the Characeae (Wood and Imahori, 1965)." The informations of these varieties and species had been presented during the Indian Botanical Society Conference at MadrasUniversity, in 1986 and National Symposium on recent trends in Marine Algal Research in India at MaduraiKamarajUniversity in 1998.

8. *Chara canescens* Desv & Lois (plate - 17)

Plants short to medium sized, 6 to 9 cm tall, rarely branched, upper main axis ecorticated; cortication starts from upper nodes of main axis and comes down; lower internodes corticated, without spine cells, 10 branchlets at a node, each 5 segmented, all segments ecorticated; sex organs at the lower nodes of branchlets. In this character, this plant differs from *C. canescens.* f. *hirsuta* R.D.W., *C. canescens* Desv. Lois, *C. canesens* Desv & Lois (*C. altaica.* A. Br.) and *C. caneseens* Desv & Lois (*C. evoluta* T.F.A.)

This is a first record of observation for Tamil Nadu on one hand and South India on the other.

9. *Chara fibrosa f. meenakshi* forma nov. D. Subramanian (Plate - 17)

The present plant was collected at Karaikudi of Pudukottai District of Tamilnadu in 1992. Holotype and Vaucher specimens are deposited at 277, Mariappa Nagar, Annamalai Nagar – 608 002, Tamil Nadu.

The plant is like fox tail and colony of plants arise from the soil in a cluster. Plants 4 to 5 cm, tall, narrow below but dilated above, unbranched; main stem corticated, 7 to 8 branchlets in a whorl, internodes shorter than branchlet whorls, each branchlet 4 to 5 segmented, all ecorticated, stipulodes larger and unequal, bracts and bracteoles longer and unequal; sex organs at the lower 2 nodes, monoecious, one antheridium and one oogonium at a node.

This plant resembles *Chara fibrosa f. keukensis* (T.F.A.). R.D.W. (Icone 131) but differs in more condensed form, absence of spine cells in the cortex of stem and 7 to 8 branchlets at a node instead of 8 to 10 in the type species.

Collection No. 245, Holotype at author's laboratory and Isotype No. 74 at B.S.I., CBE, T.N.

10. *Nitella leptostachys* A.Br.em. var. *viscosa*. D. Subramanian, 2000. Indian Hydrobiology 3(2) : 71 (Plate - 18)

Plant monoecious; to 10 to 12 cm high, axes slender, 350 μm thick; internodes slightly shorter than branchlets; primary ray 6, twice furcate, each into 3 to 4 secondary rays and each secondary ray into 2 to 3 tertiary rays, dactyl 2 celled; entire plant viscous; fertile whorls, like fruit clusters of flowering plants, arising in the axils of sterile whorls with distinct fertile heads attached in alternate pattern on either side of the stalk of fertile whorl. Oogonium 250 μm in length and 200 μm in breadth; one per each node; 8 convolutions, oospore membrane reticulate; corona cells 30 μm in length and 20 μm in breadth; antheridium quadriscutate, 200 μm in diameter, orange.

Collected by D. Subramanian at Puduchattiram, fixed in formalin and deposited under No. 175 at 277, Mariappa Nagar, Annamalai Nagar - 608 002, Annamalai University, Tamilnadu.

11. *Nitella leptostachys* A.Br.em. var. *fruiticosa*. D. Subramanian, 2000. Indian Hydrobiology, 3(2) : 71

Plant monoecious; 12 to 15 cm high; axes slender, 400 μm thick; internodes as long as branchlets, fertile and sterile whorls present. Sterile whorls viscous due to gelatinous cell wall; 6 primary laterals each into 6 secondary and each into 2 to 3 tertiary rays; dactyl 2 celled; Fertile branchlets follow the same pattern of forking of sterile in the first and second nodes; oogonium 300 μm in length and 250 μm in breadth; one per each node, 8 convolutions, oospore membrane reticulate, corona cells 40 μm in length and 30 μm in breadth; antheridium quadriscutate, 300 μm in diameter, orange.

Collected by Dr. D. Subramanian at Puduchattiram fixed in formalin and deposited under No. 176 at 277, Mariappa Nagar, Annamalai Nagar - 608 002, Annamalai University, Tamilnadu.

12. *Nitella verticillata* (Fi. & G.O.A.ex. Fil.) R.D.W. var. Kodaikanalensis, D. Subramanian, 2000. Indian Hydrobiology, 3(2) : 72 (Plate - 19)

Plant monoecious, short, 6 to 8 cm high, stem thick, to 900 μm in diameter; branchlets as long as internodes; primary rays 5, 1/2 of the length of the entire branchlet length; secondary rays 2 to 3 of which one or rarely two again fork into 3 to 4 dactyls; dactyl 2 celled; the tip cell blunt and short; upper portions of the plant produce a large number of axillary branchlets at each node, oogonium 650 μ in length, 500 μm in breadth, brown; antheridium 500 μ in diameter, orange – brown.

Collected by Dr. D. Subramanian at Fairy Falls of Kodaikanal, fixed in formalin and deposited under No. 196 at 277, Mariappa Nagar, Annamalai Nagar – 608 002, Annamalai University, Tamilnadu.

Discussion

In thallus organization with comparatively longer secondary and tertiary rays and the whole plant enclosed with mucus clouds, the first plant more or less resembles *Nitella leptostachys* A.Br.em. Unlike *N. leptostachys* which has considerably small fertile clusters one above the other in bedded arrangement, the present species has sex organs in the upper branchlet whorls of only one type. Like *N. leptostachys,* this species is also monoecious. Therefore, this species is considered as a variety of *N. leptostachys* and named herein as *N. leptostachys* var. *viscosa* var. nov. This is a rare species collected in the shallow ditches near sand dunes of Puduchattiram in paddy fields.

Another taxon collected near Portonovo also is at sandy water drains, has slightly mucus envelope and two types of whorls sterile and fertile, unlike, *N. leptostachys*. Therefore, this taxon has been considered as a variety of *N. leptostachys* and named *N. leptostachys* var. *fruiticosa* var. nov. Like the former taxon, this variety is also of vary rare occurrence.

The third taxon collected at Kodaikanal of Western Ghats, more or less resembles *N. verticillata* (Fil & G.O.A. ex Fil) R.D.W. The secondary and tertiary branchlets are arranged in verticillate pattern like those of *N. verticillata.* This verticillate arrangement of thallus is common among the species of *Tolypella*. But, because of the arrangement of sex organs which resembles those of species of *Nitella*, this taxon has been considered as a variety of *N. verticillata* and named as *N. verticillata* var. *kodaikanalensis* var. nov. This taxon was collected at Fairy Fall of Kodaikanal in abundance in 1977 and 1980 and the sex organs had been rarely observed and this is the first record in this region.

13. *Chara fibrosa* Ag. Ex. Bruz. em. *f. robusta forma* nov. D. Subramanian (Plate - 20)

This plant resembles *Chara fibrosa* of Wood and Imahori (Icone 128). Plants dioecious, 10 to 12 cm tall, robust, axes slender, 600 to 620 μm in diameter ; branched, internodes 1 ½ times as long as branchlet whorls, cortex corticated, spine cells sharp, solitary, facing downwards, stipulodes in 1 tier, branchlets 8 in a whorl, segments 5, all ecorticated, basal segment short; Bracts longer, sex organs at all nodes of branchlets, excepting the tip, inmaleplant one antheridium in each node and in female plant one oogonium in each node. This is a first dioecious member of *Chara*, the author has studied from India. Even among the forms and varieties of *C. fibrosa* studied *C. fibrosa* (*C. preissii* A.Br.). (Icone 128 of Wood and Imahori, 1975) alone is dioecious. Therefore, the present plant is the second dioecious member of *Chara fibrosa* Ag. Ex. Bruz of the world.

This form of *C. fibrosa* studied herein, differs from that of Wood and Imahori, in plant morphology, branching pattern, shorter basal segment of branchlet and presence of sex organs.

Herbarium specimens of this plant are deposited at 277, Mariappa Nagar, Annamalai Nagar – 608 002, Annamalai University, Tamil Nadu. Isotype No. 74 at B.S.I., CBE, T.N.

So far, the Western and Eastern Ghats of Tamilnadu have not been studied in detail as far as Charophytes are concerned. The Western Ghats of Tamilnadu have Ooty, Kodaikanal and Valparai as more important hill stations. The Eastern Ghats of Tamilnadu have Kanyakumari and Thirunelveli Hills, Kolli Hills, Kalrayan and Silver hills, Shervarayan hills, Pachamalai Hills, Gingee hills etc. Eventhough the elevations are less than 6000 or 5000 feet, luxurious growth of various groups of plants have been observed in these areas.

Yercaud is the capital of Shervarayan hills consisting of 40 to 50 villages. The elevation of this is 5000 feet and floristically not investigated thoroughly. A robust form of *Chara* has been collected from Yercaud and the Herbarium specimen deposited at author's laboratory at 277, first South Cross Road, Mariappa Nagar, Annamalai Nagar – 608 002, Tamil Nadu.

14. *Nitella grandiformis*. var. *hyalinoides* sp. et. var. nov. D. Subramanian (Plate - 21)

This plant has more or less the same characters of *N. grandiformis* var. *lhotskeyiana* sp. et. var. nov. already described by the author (Plate 11). The important differences are plant size and morphology and one of the dactyls of secondary rays converted into tertiary ray, ending into again 2 daclyls. As this plant resembles *N. hyalina* it is named herein as var. *hyalinioides*.

Holotype at author's laboratory and Isotype at B.S.I., CBE, T.N.

15. *Chara vulgaris* var. *yercaudensis* var. nov. D. Subramanian (Plate - 22)

Plants 7 to 9 cm tall, heavily incrusted, a robust plant, internodes 1 1/2 times longer than branchlet whorls; 10 branchlets in each whorl; branchlet 9 to 10 segmented; segments bulged and shorter, bracts and bracteoles short, no spine cells; stipulodes absent, sex organs not seen.

In the nature of segments, this plant resembles *C. vulgaris* var. *kirghisorum f. hungarica* (fi.) R.D.W. (Icone 19). But in all other characters this plant is different from the above form of *C. vulgaris*. Besides, this present plant is not at all comparable to any forms and varieties of 31 plants described by Wood and Imahori (1965) in *C. vulgaris*. So also this present plant is not at all related to any member of *Chara* described by the above authors in their book. Therefore, the present plant is very interesting and the nature of segments of branchlets and other characters of the plant can only be observed under 20 x lens of Dissection microscope and compound microscope with 100 x or 400 x or above magnifications will not help to study morphological characters of the plant. This new variety is named here as *yercaudensis* var. nov. But, unfortunately only one plant is available in the herbarium sheet and the author will try to find some more plants with sex organs.

Top slip is one of the more attractive pilgrimage centres and Tiger sanctuary in Coimbatore Hills. It is one of the hills of Eastern Ghats and the elevation is 3500 feet. When the author went on excursion in 1984 along with M.Sc. Botany students of AnnamalaiUniversity, he came across in a small forest stream, a species of *Nitella*. At that time (1984), the author was not knowing much about Charophyta and only in 1988, when he went to Germany (Berlin, West) to attend the International Genetics Conference, he happened to see the two volumes of book on Charophyta written by Wood and Imahori (1965).

The only plant collected by the author at Topslip was peculiar in having distinct dome like but, small fertile whorls in the axil of sterile whorls. The sterile whorls are simple, long and terminally once forked. But, the fertile whorls made up of 3 clusters one above the other, crowded, are smaller with antheridia and oogonia. Eventhough author had already seen the monograph on Indian Charophyta by Pal, Kundu, Sundaralingam and Venkatraman (1962) and a paper by V.S. Sundaralingam (1959) on South Indian Characeae, he could not see a plant like that of Topslip. Now, the author has 2 herbarium specimens of that plant and drawings of 1984. It is a rare species of *Nitella* described below:

16. *Nitella arechavaletae var.* topslippensis var. nov. D. Subramanian (Plate - 23)

Plants monoecious, 10 to 12 cm. high, freely branched, fertile and sterile whorls separate, sterile whorls, 5 to 6 branchlets, each branchlet once forked, dactyls 2 celled; fertile whorls axillary, much reduced in size but 3 to 4 whorls one above the other, fertile whorl with 5 branchlets, each branchlet one or two times forked, dactyls 4 or 5 celled; at each node either two oogonia or one oogonium and one antheridium; 270 to 300 μm in length; 220 to 230 μm in breadth, 10 or 11 convolutions, Antheridia small, 120 to 150 μm in diameter.

The present plant is shorter and more branched when compared to type species *N. arechavaletae* speg. described by Wood and Imahori (1964) in their book (Icone No. 369). Further, in the number of branchlets and nature of furcations of them in both fertile whorls and sterile whorls, the present plant is different from that of Wood and Imahori.

Isotype kept at author's home laboratory at Mariappa Nagar, Annamalai Nagar, T.N, and Holotype deposited at B.S.I, CBE, T.N.

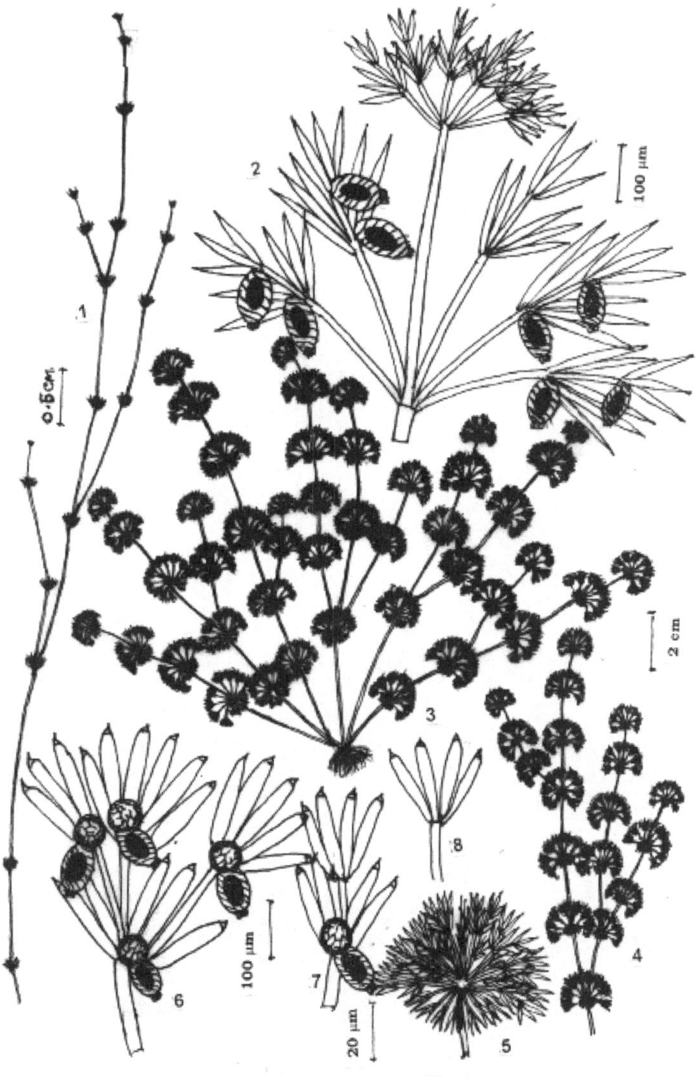

Plate 10 *Nitella congesta var filiformis* Figs 1 & 2. *Nitella congesta var. filiformis*. 1 = A Plant; 2 = A whorl enlarged. Figs. 3 to 8: *Nitella hyalina*. 3 & 4 = Plants; 5 = A whorl enlarged; 6 to 8 = 3 types of branchlets.

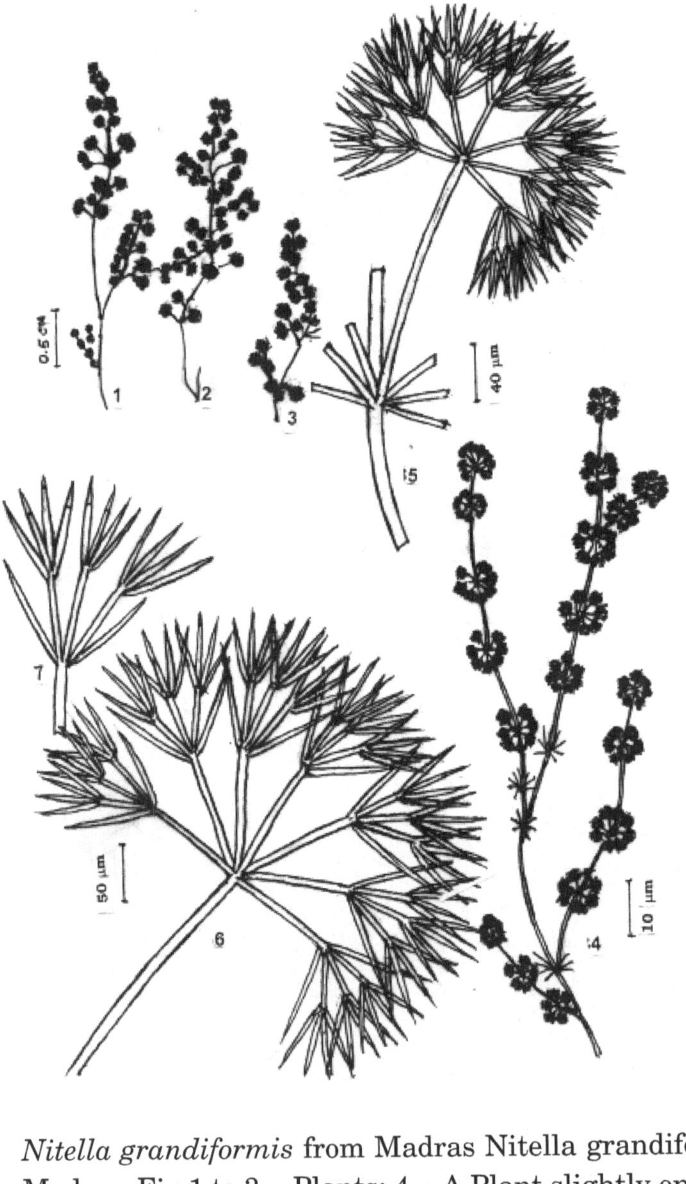

Plate 11 *Nitella grandiformis* from Madras Nitella grandiformis from Madras. Fig 1 to 3 = Plants; 4 = A Plant slightly enlarged; 5 = Axial node with branchlets; 6 = A branchlet enlarged; 7 = tip of branchlet.

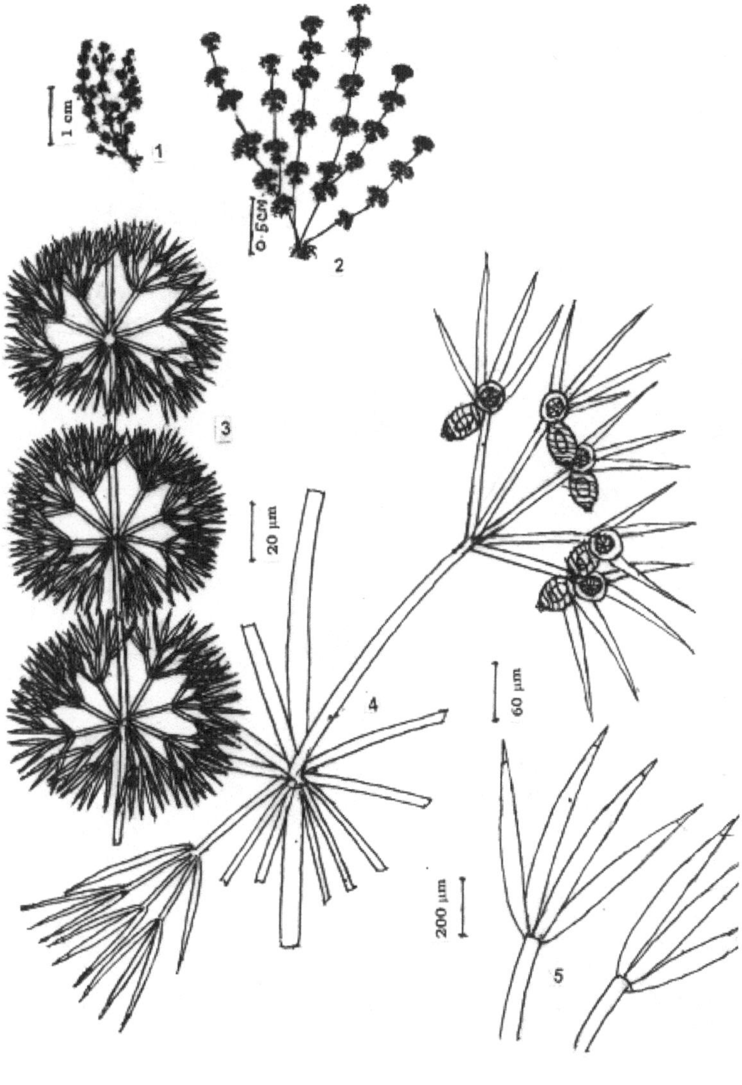

Plate 12 *Nitella lhotskeyi f. hyalinoides* Fig 1 = Plant natural size; 2 = plant slightly enlarged; 3 = A portion enlarged; 4 = A whorl with a large branchlet and a small branchlet; 5 = dactyls.

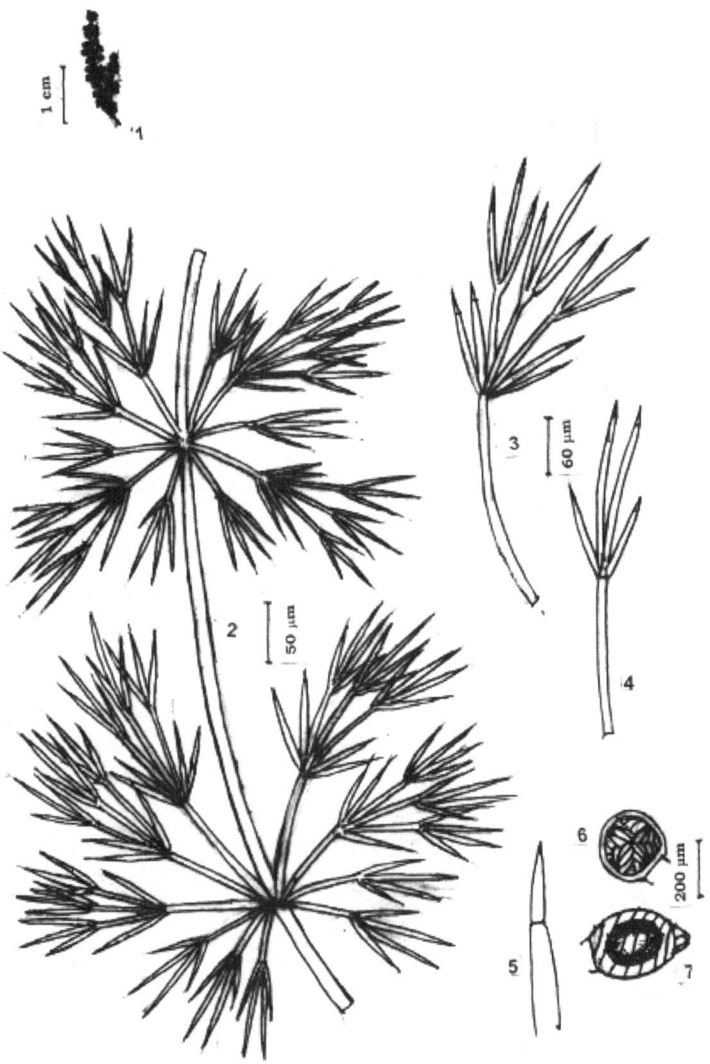

Plate 13 *Nitella lhotskeyi f. minima* Fig 1 = A plant, natural size; 2 = 2 whorls enlarged; 3 = large branchlet; 4 = small branchlet; 5 = A dactyl; 6 = Antheridium; 7 = Oogonium.

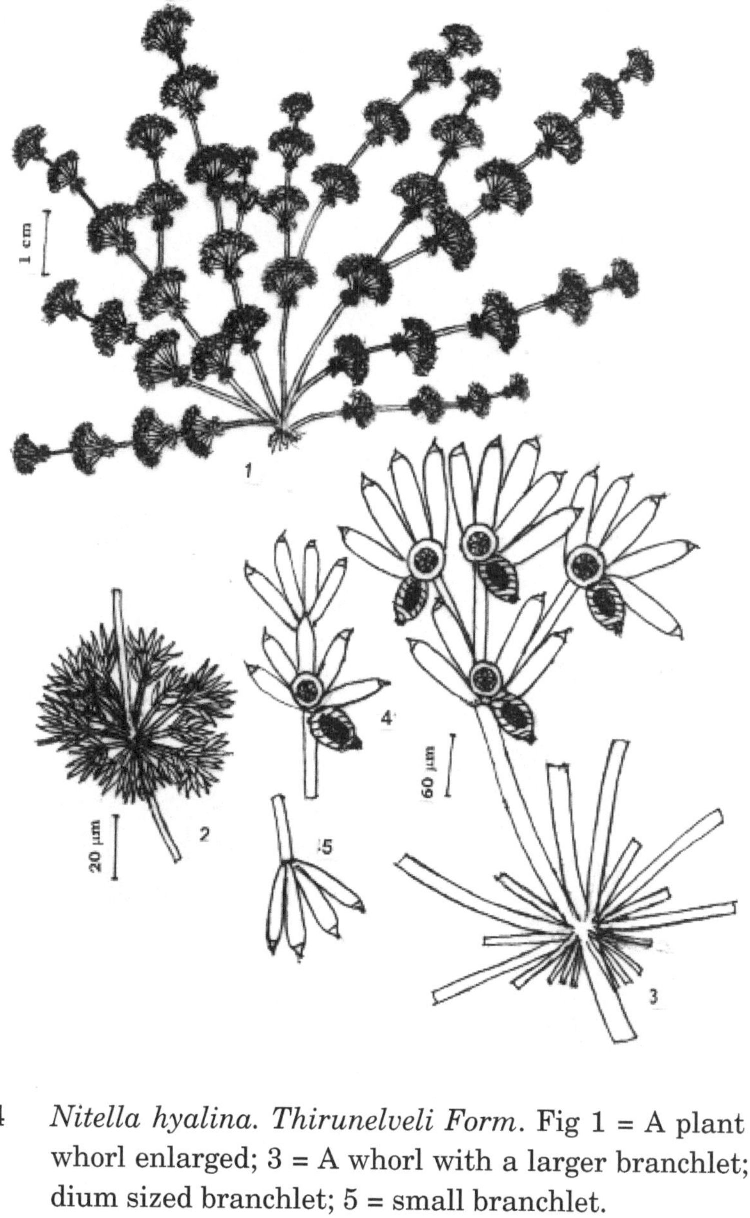

Plate 14 *Nitella hyalina. Thirunelveli Form.* Fig 1 = A plant = 2 = A whorl enlarged; 3 = A whorl with a larger branchlet; 4 = medium sized branchlet; 5 = small branchlet.

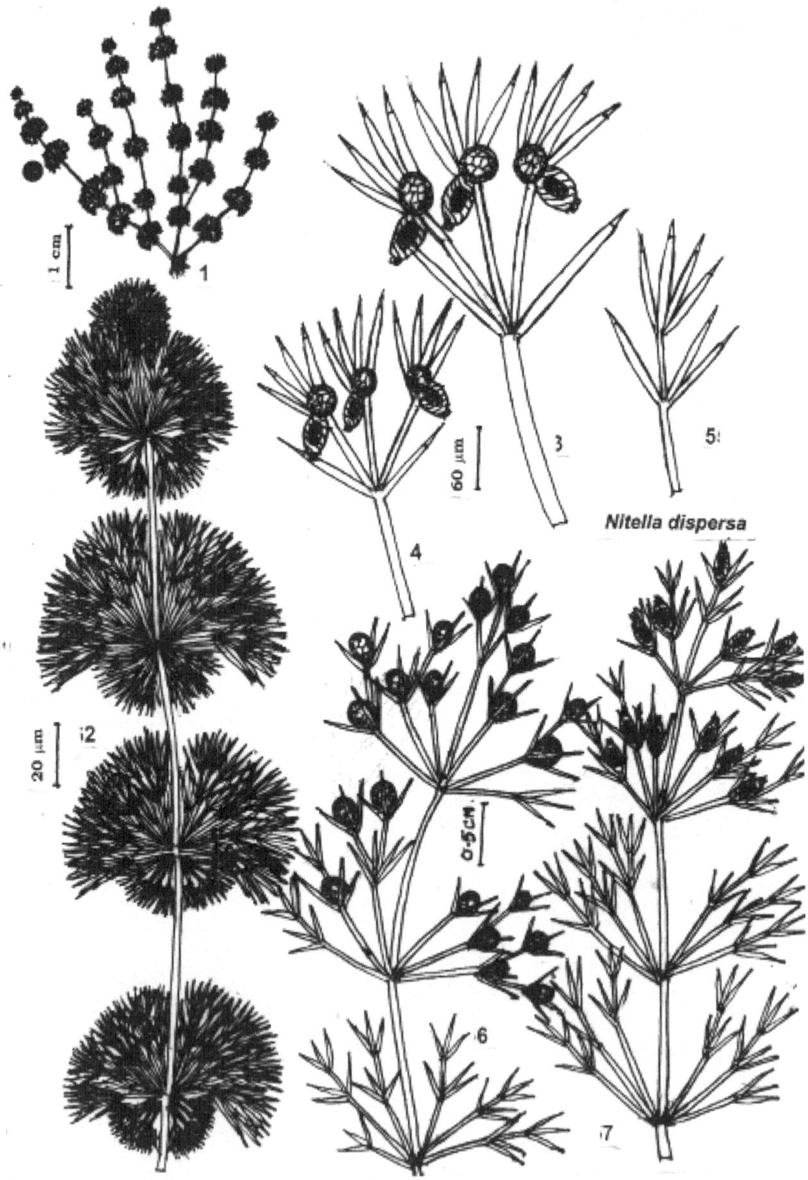

Plate 15 *Nitella hyalina*. Figs. 1 to 5 = Mavidanthal Form. Fig. 1 = A Plant; 2 = Portion of plant enlarged; 3 to 5 = 3 types of branchelts. Figs 6 and 7 = *Nitella dispersa*. Mavidanthal Form. Fig. 6 = Male Plant; 7 = Female Plant.

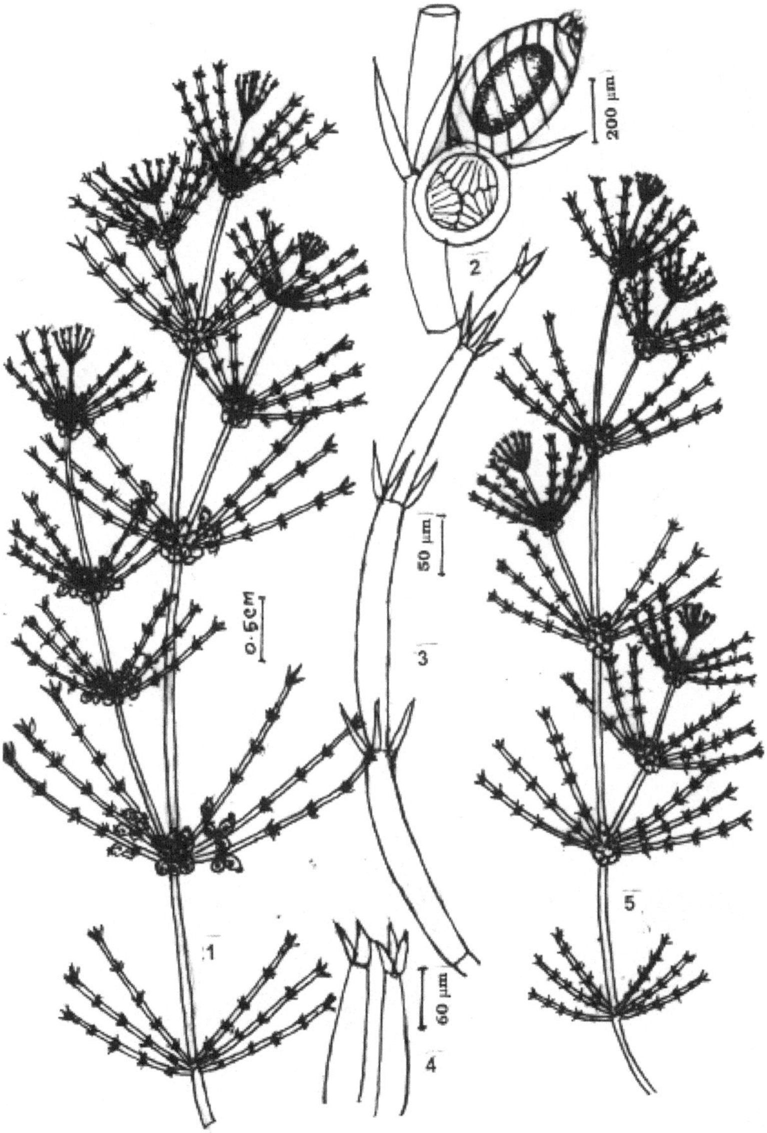

Plate 16 *Chara corallina* Figs 1 to 4 = Form 1. Fig 1 = A Plant; 2 = Sex Organs; 3 = A branchlet; 4 = branchlet tip. Fig 5 = Form 2. A.Plant

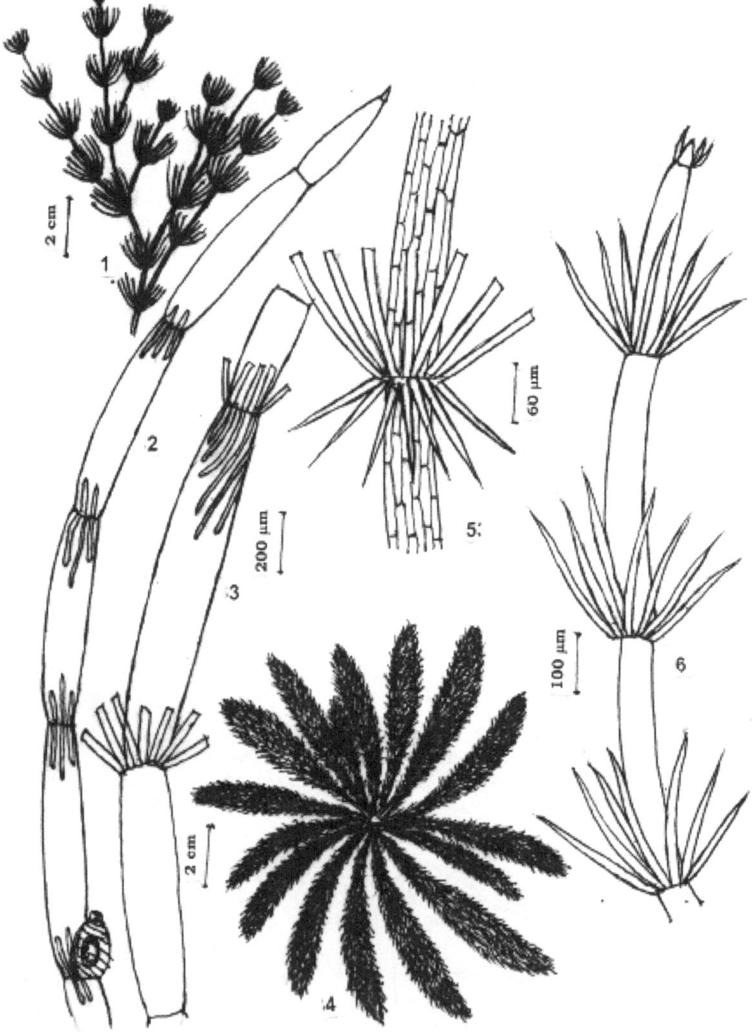

Plate 17 *Chara canescens* Figs 1 to 3 = Chara canescens. Fig. 1 = A plant; 2 = A branchlet; 3 = Two nodes, one with A wining cortical threads and another with branchlets. Figs. 4 to 6 = *Chara fibrosa f.meenakshi*. Fig 4 = a plant; 5 = Axis node; 6 = A branchlet.

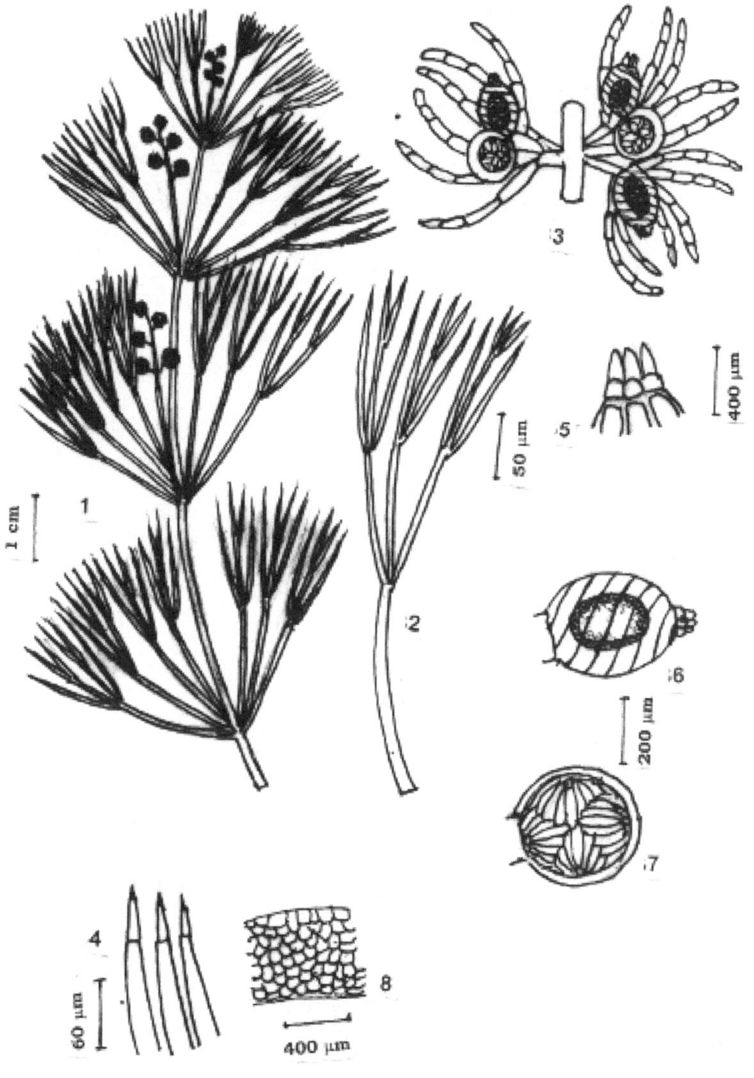

Plate 18 *Nitella leptostachys* var. *viscosa Puduchattiram type* Fig. 1 = A plant enlarged; 2; sterile branchlet; 3 = Fertile whorl; 4 = Dactyls = 5 = Corona cells; 6 = Oogonium; 7 = Antheridium; 8 = Oospore membrane.

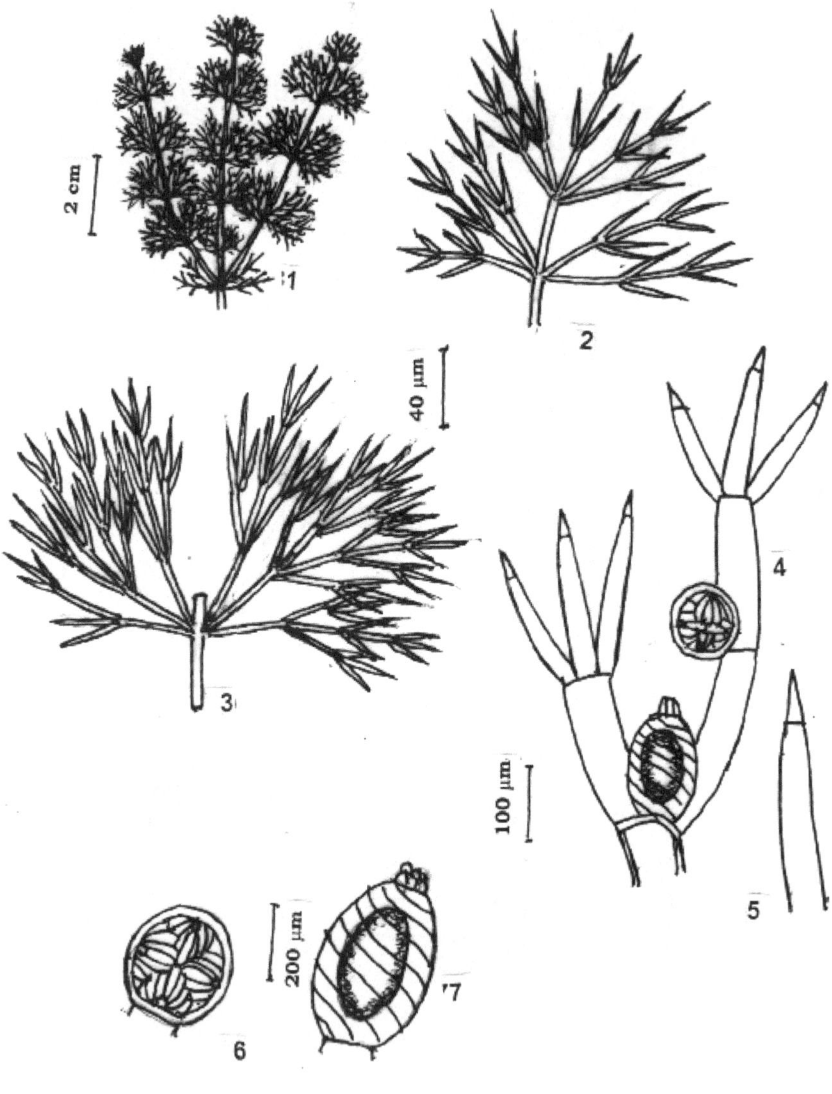

Plate 19 *Nitella verticillata*. var. *kodaikanalensis*. Fig. 1 = A plant; 2 and 3 = whorls; 4 = Sex organs; 5 = Dactyl; 6 = Antheridium; 7 = Oogonium.

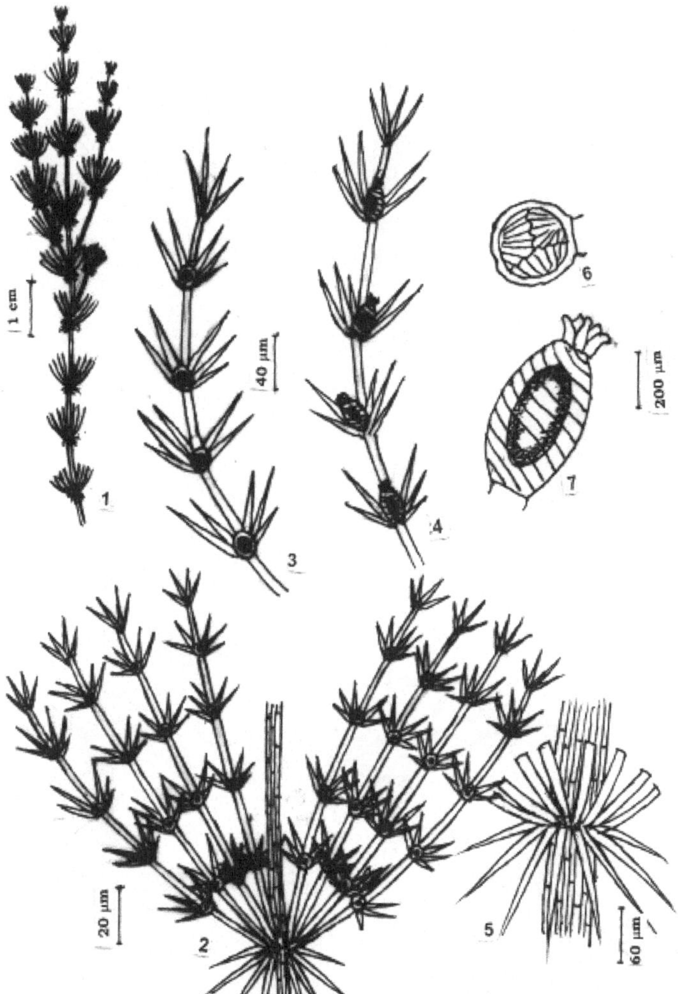

Plate 20 *Chara fibrosa. f.robusta*. Sankarapuram Form. Fig.1 = Plant; 2 = A whorl with branchlets; 3 = branchlet of male plant; 4 = branchlet of female plant; 5 = An axial node with stipulodes; b = Antheridium; 7 = Oogonium

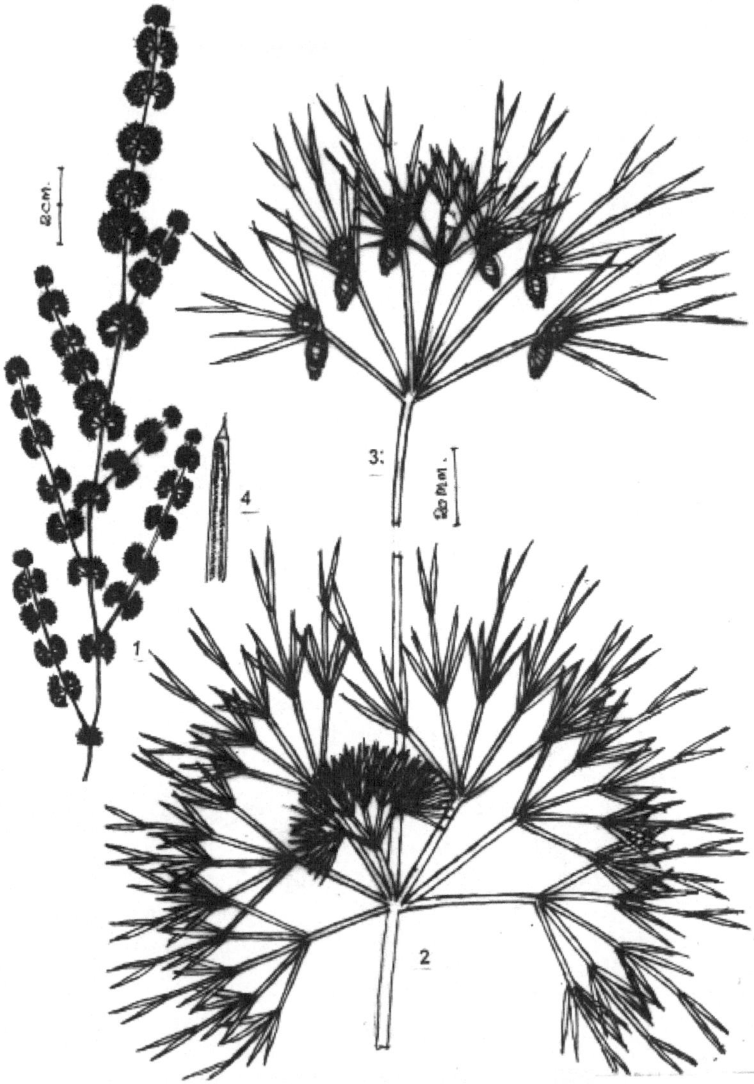

Plate 21 *Nitella grandiformis*. var. hyalinioides. Fig. 1 = A plant; 2 = A vegetative whorl with axillary fertile whorl in the axis; 3 = fertile whorl. 4 = Dactyl.

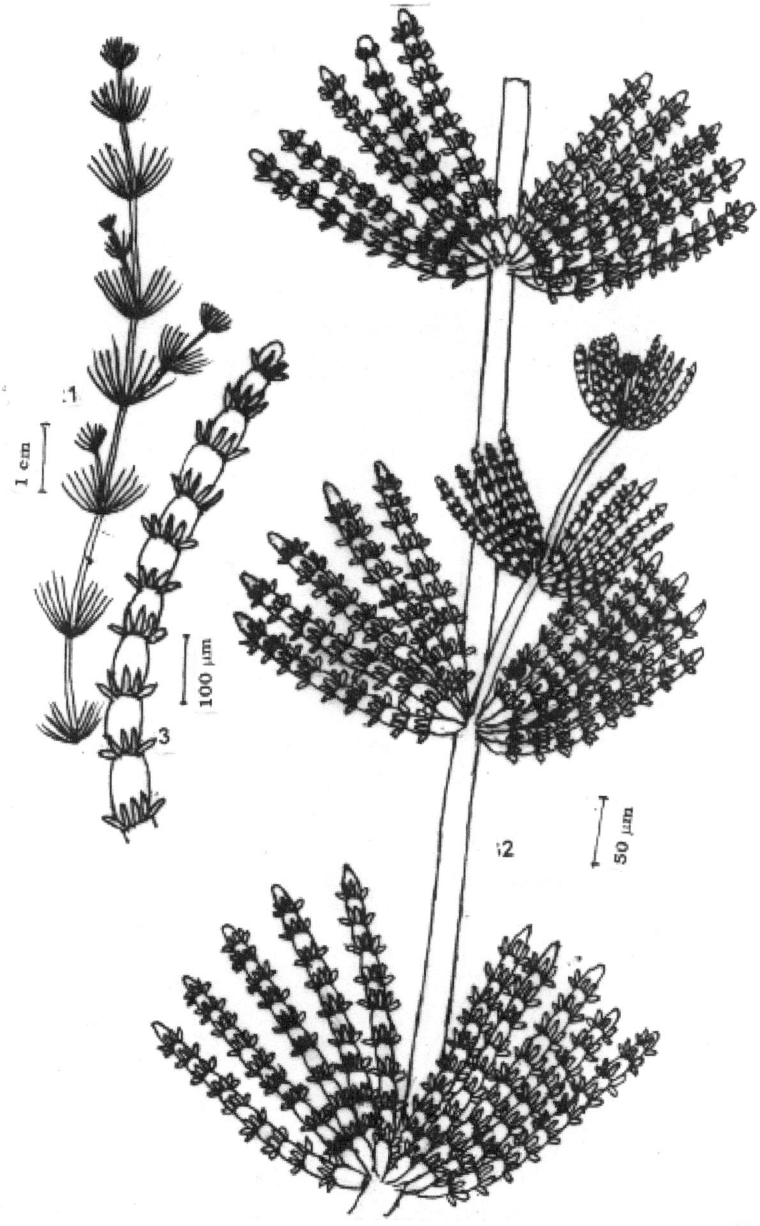

Plate 22 *Chara vulgaris*. var *yercaudensis*. Fig. 1 = A plant; 2 = A portion of plant enlarged; 3 = A branchlet.

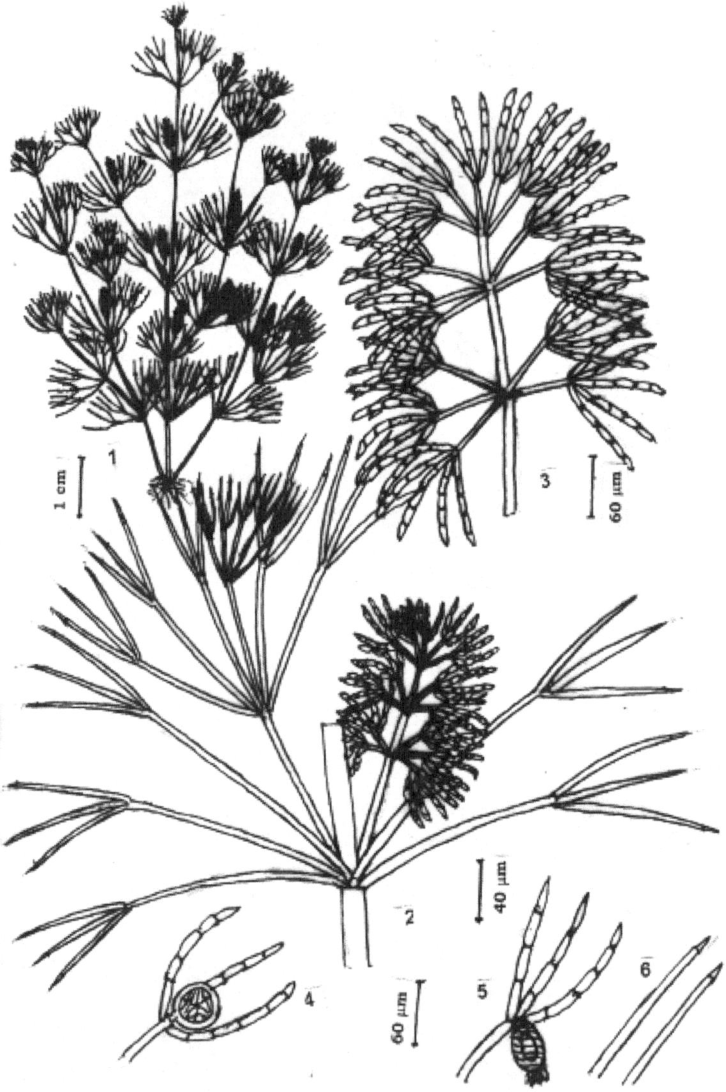

Plate 23 *Nitella arechavaletae*. From Topslip. Fig. 1 = A plant; 2 = A sterlie whorl with an axillary fertile whorl; 3 = Fertile whorl enlarged; 4 = Antheridium; 5 = Oogonium; 6 = Dactyls.

3. CHAROPHYTIC MEMBERS FROM THIRUVANNAMALAI OF TAMILNADU

There is a large lake on the way to Thiruvannamalai from Thirukkovilur of Tamilnadu by the side of bus route. The author could collect peculiar varieties and forms of *Chara fibrosa, Nitella furcata, N. translucens, N. flexilis* and *N. hyalina*. There are 2 forms in each of *N. translucens* and *N. flexilis* and 3 forms of *N. hyalina*. Even with our field observation, we could find out the differences in the morphology of the forms of these species. The author carefully examined these forms under different magnifications of compound microscope and described them.

1. *Chara fibrosa* var. *robusta* D. Subramanian var. nov. (Plate - 24).

This plant was collected at the borders of the large lake of Thiruvannamalai.Plants are robust in texture, 8 to 9 cm in height, medium green, main stem provided with downwardly pointing hook like bristles, stipules and stipulates are long and spine like, 12 branchlets, 24 bracteoles; all the 5 segments of branchlet ecorticated, sex organs at the basal 3 nodes only, 2 to 3 antheridia and 2 oogonia at each node; antheridium Orange and oogonium brown.

Herbarium specimens are deposited at author's laboratory at Mariappanagar, Annamalainagar-608 002, Tamilnadu, Collection No. 386, dated 20.02.2006. Isotype No. 13 at B.S.I., CBE, T.N.

2. *Nitella furcata* subsp. *mucronata* R.D.W. (Plate - 25)

This plant closely resembles to the type species as described by Wood and Imahori.Plants 8 to 10 cm tall, rarely branched; 6 branchlets at each node, the dactyls are 2 or 3 celled and mucronate. This is one of the commonest members of Charophytes collected at Thiruvannamalai lake on 20.02.2006 collection No. 389. Isotype No. 18 at B.S.I., CBE, T.N.

3. *Nitella translucens* (Pers) Ag. Em. var. *minima* var. nov. D. Subramanian (Plate -26)

This is a small species, 6 to 7 cm tall; basal sterile whorls 3 times furcate, tip whorls of unequal rays, dactyl one celled, short and acuminate. At each node of fertile whorls 6 primary rays present, each ending into 4 secondary rays, almost equal; in each node one antheridium above and one or two oogonia by the sides. This is a simple variety of this species, the author has seen so far.

Plants collected at Thiruvannamalai lake on 20.02.2006 Collection No. 403. Herbarium specimens are deposited at author's laboratory at Mariappa Nagar, Annamalai Nagar–608002, Tamilnadu.

Isotype No. 12 at B.S.I., CBE, T.N.

4. *Nitella translucens* (Pers.) Ag. em var. *annamalaiensis* var. nov. D. Subramanian (Plate - 27)

This variety is larger than the first variety of this species, 7 to 9 cm tall, mostly unbranched; fertile branchlets clearly demarked from the sterile ones on either side of them.

Holotype at author's laboratory; Isotype No. 76 at B.S.I., CBE, T.N.

5. *Nitella flexilis* (Pers.) Ag. em. var. *lancifolia* var. nov. D. Subramanian (Plate - 28) Collection No. 405; dated : 20.02.2006.

This is larger than the previous plant, 9 to 10 cm tall, rarely branched; at each node of main stem, 6 sterile branchlets and 1 or 2 or even 3 fertile branches arise; the spreading of branchlets both sterile and fertile are like'Hand Fans made from *Borassus flabellifer*" and very beautiful to look at.

Holotype at author's laboratory; Isotype No. 77 at B.S.I., CBE, T.N.

6. *Nitella flexilis* var. *grandifolia* var. nov. D. Subramanian (Plate - 29) Collection No. 407; Dated : 20.02.2006.

This variety of *N. flexilis* is larger in size and more in number of branches, branchlet whorls, both sterile and fertile whorls are beautifully arranged, when we see this plant under low power of compound microscope.

Both this and previous variety of *N. flexilis* were collected in 1 to 2 feet depth of water near the borders of the lake near Thiruvannamalai. Voucher herbarium specimens are deposited at author's laboratory at Mariappanagar, Annamalainagar – 608 002, Tamilnadu.

Isotype No. 78 at B.S.I., CBE, T.N.

7. There are 3 distinct forms of *Nitella hyalina*. Collection Numbers 410 to 412)

There are 3 distinct forms of *Nitella hyalina* more or less in the same locality of the lake at 1 1/2 to 2 feet depth of water. These members are very attractive when we see them in their natural conditions.

These forms (Plates 30 to 32) are distinct mainly in the nature of final rays ending into dactyls. In the first form, the tip rays are almost equal, narrowly elongated, conical and needle like. In the second form, the final rays narrow rectangular ending bluntly into dactyls and all of them are equal in size.

In the third form, although the final rays resemble those of first form, they are broadly conical and more or less unequal. The entire whorls of the first form are encrusted with mucilage clouds but in the second and third forms, the distinct whorls are not covered by mucilage clouds. The presence of 3 forms of *N. hyalina* is very curious in one and the same place. The soil is reddish clay and the water is clean like glasses. So far, this region of Tamilnadu has not been studied not only for Charophytes but also for other groups of fresh water algae. Like this, there are so many places of peculiar ecological environmental conditions, where no study in fresh water algae has been made so far.

The forms, varieties and species of the *Chara* and *Nitella* observed at Thiruvannamalai lake are almost distinct,when compared to those studied from Cuddalore and Tanjore Districts.

The soil type, latitude and temperature play important role in plant morphology of Charophyta and this study is an example for the same. In most of the areas of Cuddalore District the soil is clayish or sandy loam and they are near sea shore of Bay of Bengal and mostly not so hot. But the places of Thiruvannamalai District are interior on the way to Bangalore from Tamil Nadu, where there is generally hot climatic conditions and mostly red or loamy soil.

These environmental and soil conditions are responsible for the deviant morphological conditions of *Chara* and *Nitella* in these two places.

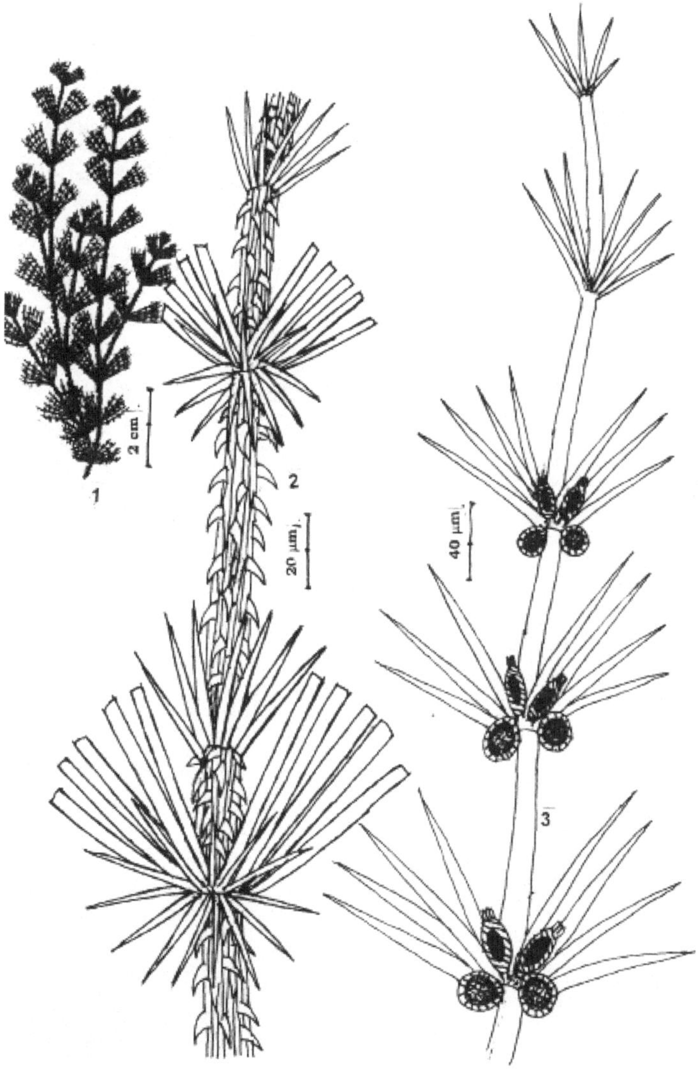

Plate 24 *Chara fibrosa*. var. *robusta*. Fig. 1 = A plant; 2 = Main axis showing stipulodes, spine cells and branchlets; 3 = A branchlet with sex organs.

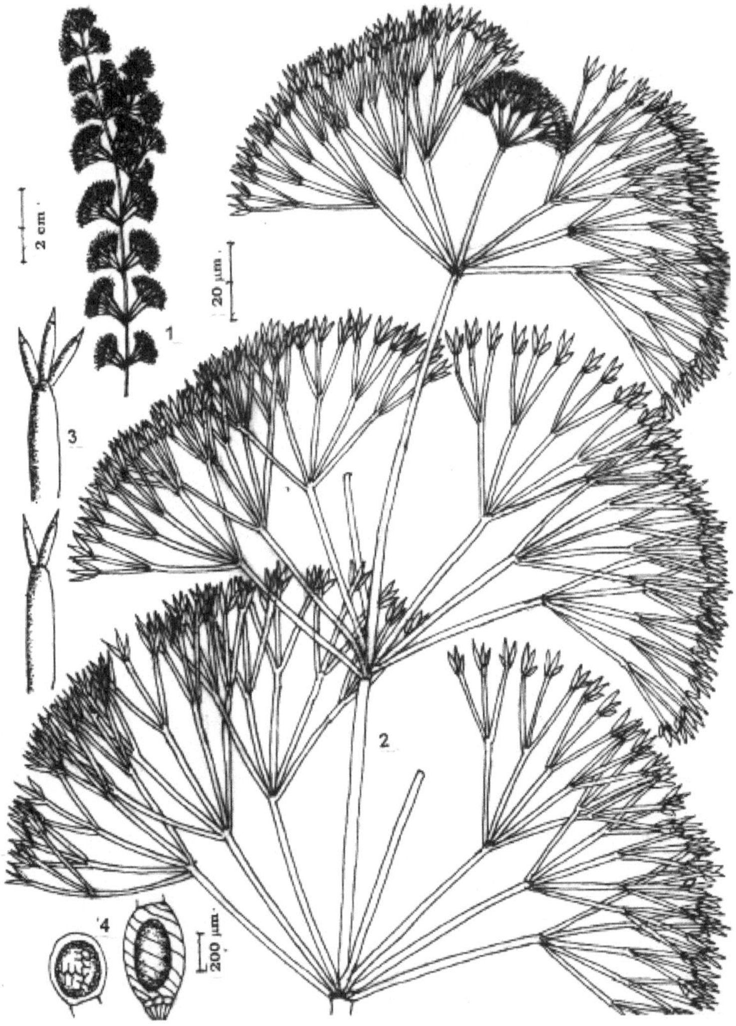

Plate 25 *Nitella furcata* subsp. *mucronata*. Fig. 1 = A plant; 2 = A portion of the plant enlarged; 3 = dactyls; 4 = sex organs

Plate 26 *Nitella translucens* var. *minima*. Fig. 1 = A plant; 2 = Basal sterile whorl with a branchlet; 3 upper fertiler whorls; 4 = Dactyl;

Plate 27 *N. transluecens* var. *annamalaiensis*. Fig. 1 = A plant; 2 = portion of a plant with sterile whorls and fertile whorls enlarged; 3 = Dactyls; 4 = Oogonium; 5 = Antheridium.

Plate 28 *Nitella flexilis*. var. *lancifolia*. Fig. 1 = A plant; 2 = A portion of plant with sterile and fertile whorls enlarged; 3 = A dactyl.

Plate 29 *N. flexilis*. var. *grandifolia*. Fig. 1 = A plant; 2 = A portion of a plant with sterile and fertile whorls enlarged.

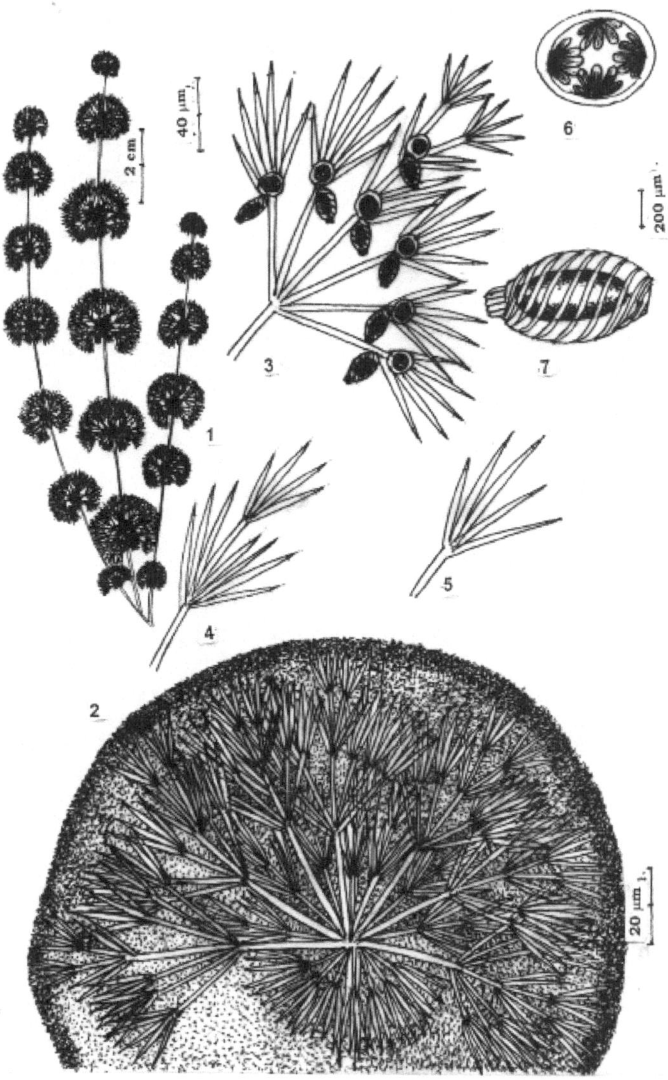

Plate 30 *Nitella hyalina*. Form 1. Fig. 1 = A plant; 2 = A whorl within mucilage clouds; 3 to 5 = 3 types of branchlets; 6 = Antheridium; 7 = Oogonium.

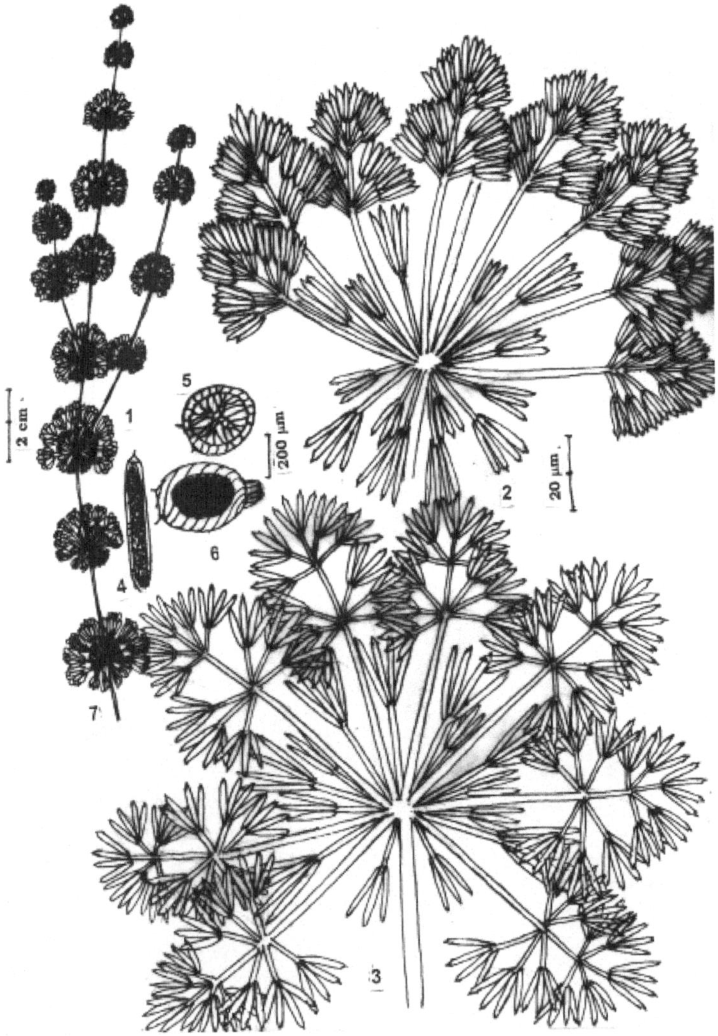

Plate 31 *N.hyalina*. Form – 2 Fig. 1 = A plant; 2 & 3 = 2 whorls enlarged; 4 = Dactyl; 5 = Antheridium; 6 = Oogonium.

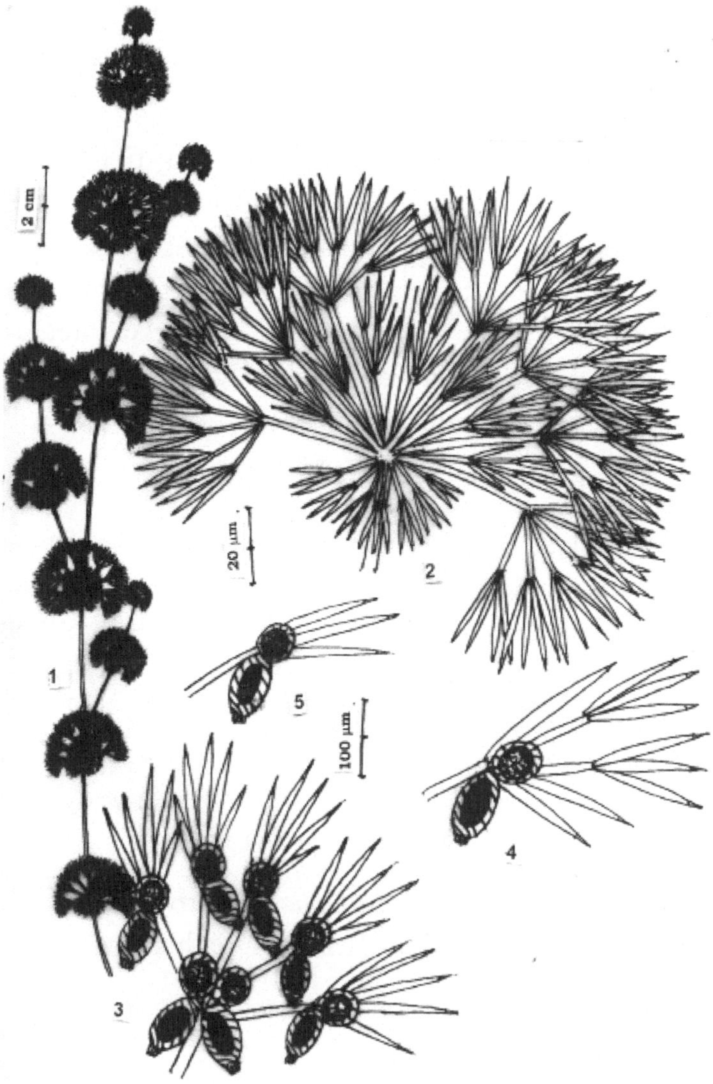

Plate 32 *N.hyalina*. Form – 3 Fig. 1 = A plant; 2 = A whorl enlarged; 3 to 5 = 3 types of branchlets.

4. CHAROPHYTIC FLORA OF VIRUDHACHALAM TALUK OF TAMIL NADU

1. ***Chara zeylanica* Kl ex Willd. Type -1 (Collection No. 235 dated 05.01.2000 at Ko Athanur)**

   Plants forming medium green coloured dense mat; 10 to 12 cm in height, profusely branched; at each node 10 branchlets, each branchlet with 10 segments; all corticated excepting basal shorter segment which is ecorticate; sex organs borne at first to third nodes only with unequal larger bracts; all the other nodes with shorter bracts; antheridium large, 800 to 900 μm in diameter, orange, quadriscutate; oogonium large, 850 to 900 μm long and 450 to 500 μm broad; light brown; 5 coronula cells, prominent.

   Type II (Collection No. 236 dated 05.01.2000 at Ko Athanur).

   Plants medium sized; 8 to 10 cm in height; isolated, dark green and rarely branched; at each node 12 branchlets, each branchlet with 9 segments, tip segment broad below and narrow above; all segments corticated; bract cells at the lowest nodes, shorter than sex organs and rudimentary at the upper most nodes of branchlets; sex organs on 1 to 3 or 4 nodes; antheridium small, 200 to 250 μm in diameter, brown, quadriscutate; oogonium small, flask shaped; brownish black, 400 to 450 μm long and 300 to 350 μm broad with 5 shorter and broader coronula cells.

2. ***Nitella flexilis*** (L.) Ag., em. var. ***Sathamangalensis*** var. nov. D. Subramanian (Collection No. 237 dated 06.01.2000 at Sathamangalam).

   Plants monoecious, up to 20 cm high, much crowded at the tip of the plant; fertile and sterile whorls present, sterile whorls 7 to 8 per node, once furcated into 3 dactyls, dactyl one celled, fertile whorls with 2 to 3 nodes, arrangement of branchlets of fertile whorls follows the same pattern as sterile whorls. Fertile whorls comparatively smaller than sterile whorls, 2 to 3 fertile whorls per node, sex organs at the tip of each primary rays; oogonium medium sized, 400 to 450 μm long and 300 to 320 μm broad, orange with 10 coronula cells and 9 spiral cell convolutions; oospore membrane granulate to closely papillate; antheridium orange, 400 to 420 μm in diameter, octascutate.Holotype at author's laboratory and Isotype No. 44 at B.S.I., CBE, TN.

3. ***Nitella annamalaiensis*** **D. Subramanian Phykos. 39 (1&2) : 36, 2000 (Collection No. 238 dated 06.01.2000 at Sathamangalam; Plate - 33)**

Plants monoecious, thick green and short, 10 cm high; primary rays at each node more than ½ of the entire branchlet length, each primary ray furcates into 3 to 4 secondary rays, each secondary ray forking into 3 to 4 unequal dactyls, dactyl 2 celled, tip cell sharp and prolonged; fertile whorls at the first or rarely second branchlet nodes, smaller and following the same pattern of forking as sterile whorls; at each node one oogonium and one antheridium; oogonium, brown, 500 to 550 μm longand 400 to 420 μm broad; 10 corona cells; oospore membrane vermiformis; antheridium orange,350 to 380 μm in diameter, octascutate.Isotype at author's laboratory; Holotype No. 79 at B.S.I., CBE, T.N.

4. ***N. verticillata*** **(Fil. & G.O.A. ex Fil) R.D.W. (Collection No. 239; dated 06.01.2000 at Sathamangalam; Plate - 34)**

A small species of 5 to 6 cm height, 6 branchlets in a whorl, each primary ray ½ of the length of entire branchlet, forking into 2 dactyls and 1 secondary ray ending into one or two unequal dactyls, 1 celled, verticillate arrangement, sex organs not found. Holotype at author's laboratory; Isotype No. 68 at B.S.I., CBE, T.N.

5. ***N. doidgeae*** **J. Gr. & Steph. (Collection No. 240 ; dated 06.01.2000 at Sathamangalam; Plate - 35)**

Plant 8 to 10 cm high, light green, crowded at the tip, monoecious; terminal branchlets longer than internodes; 6 primary rays at a node, 1/3 of entire branchlet length; primary ray into 4 secondary rays, each secondary ray into 2 to 3 dactyls or 4 tertiary rays, each tertiary ray into 2 dactyls, dactyl 2 celled, lower cell bulged and upper cell tapering into a point; in the first node, 2 oogonia and one antheridium or one antheridium or one oogonium present, oogonium orange green; globular; 350 to 400 μm long and 250 to 300 μm broad; 10 coronula cells, 8 spiral cell convolutions; antheridium light orange; 400 to 450 μm in diameter; octascutate.

Nitella tenuissima (Plate – 35a) and *N. pseudoflagellata* were also recorded in the study conducted by the author. These species differ in a few morphological characters like nature of furcations, size and shape of sex organs, dactyls and general habit of the plants from the type specimens described by Wood and Imahori (1965).

Discussion

Nitella doidgeae is a first record for India and for Tamilnadu. Wood and Imahori (1965) have described a specimen of this species but it is dioecious and in the female plant only immature oogonia have been observed and in the male plant no antheridium has been noted. In all other morphological characters, the present specimen and that of Wood and Imahori from Africa are more or less similar. No mucus was observed in the specimen studied and the plants were slightly viscous. But, the specimen described by Wood and Imahori was covered by mucus. Therefore, the present specimen studied may be distinct form of *N. doidgeae*, namely *f. congesta*.

Only few specimens have been collected from the water canals coming from the natural pond in Sathamangalam, Virudhachalam Taluk, Tamil Nadu.

G.O. Allen had loaned the material of *N. verticillata* collected from Western Australia to Filarsky (1937) who published this species as *Charina verticillata* Filarsky and G.O. Allen ex. Filarsky. But the specimens were incomplete and without sex organs and had monopodial branchlets, a character of *Tolypella*. The present material has no sex organs and therefore, attempt will be made to collect this species with sex organs to ascertain the exact status.

In *Nitella myriotrichia* A. Br. ex Kittz., fertile heads arise terminally or axillarily from the main shoot node or occasionally from the first nodes of branchlets. But in *Nitella annamaliensis* proposed here, the fertile heads usually arise at the first nodes of all branchlets. In *N. myriotrchia*, sex organs may be also found rarely on strile whorls but in *N. annamalaiensis* the sex organs do not arise from sterile whorls. *N. myriotrichia* is dioecious but *N. annamalaiensis* is monocious. Therefore, this species has been treated as a new species and named *N. annamalaiensis*.

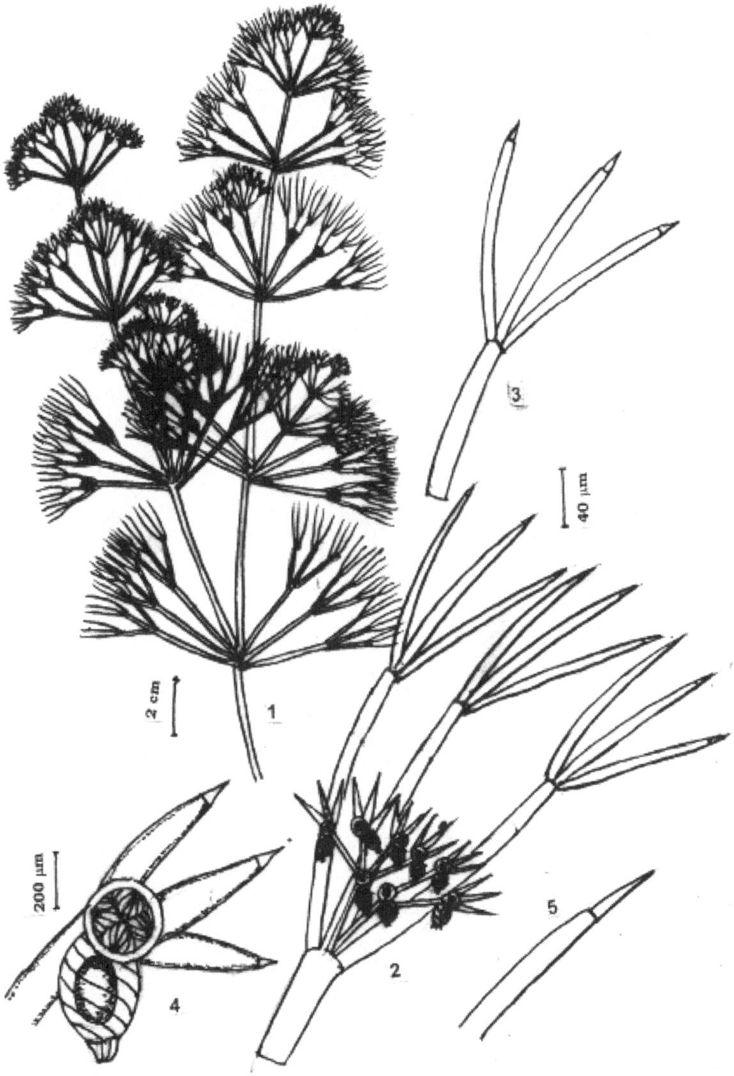

Plate 33 *Nitella annamalaiensis*. Fig. 1 = A plant enlarged; 2 = A fertile whorl arising from the first node of sterile branchlet. 3 = sterile branchlet; 4 = Fertile branchlet; 5 = A dactyl.

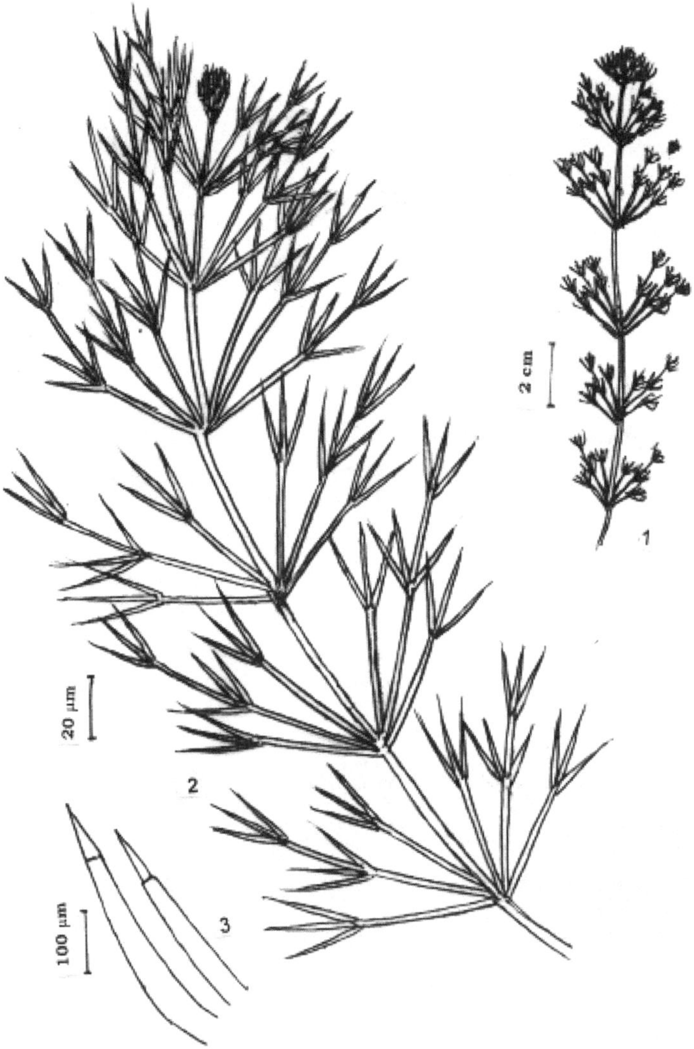

Plate 34 *Nitella verticillata*. From Sathamangalam. Fig. 1 = A plant; 2 = A plant enlarged; 3 Dactyls.

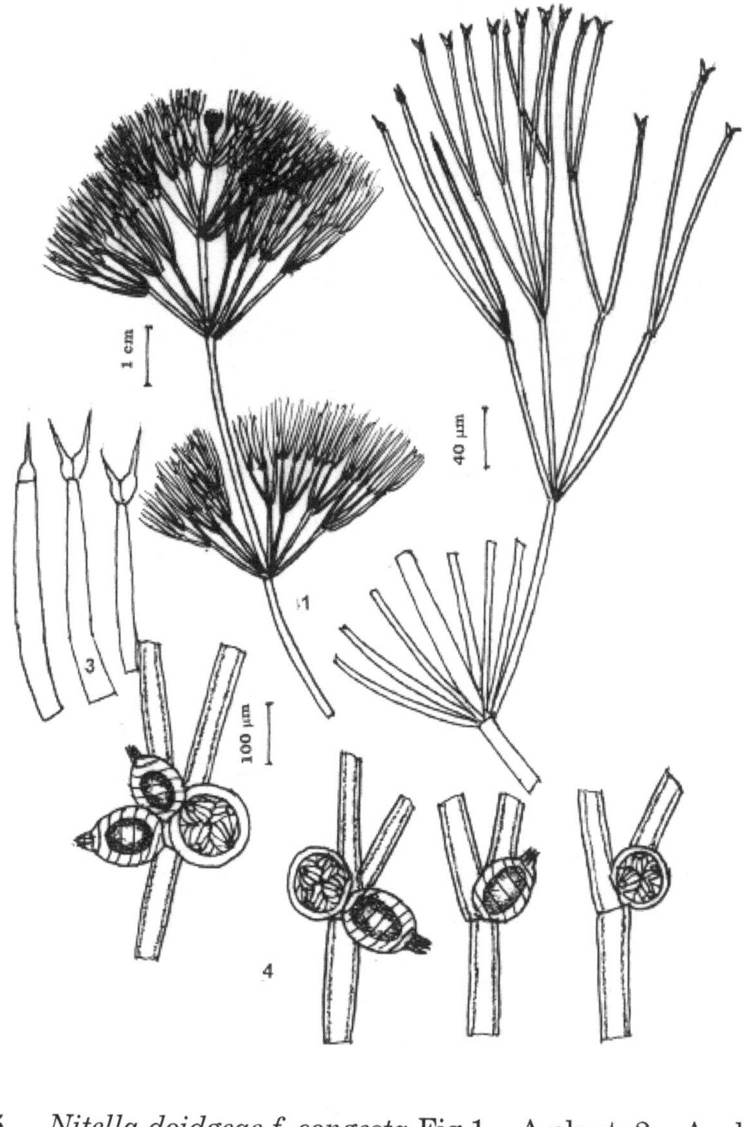

Plate 35 *Nitella doidgeae f. congesta* Fig.1 = A plant; 2 = A whorl with a branchlet enlarged; 3 = dactyls; 4 = sex organs.

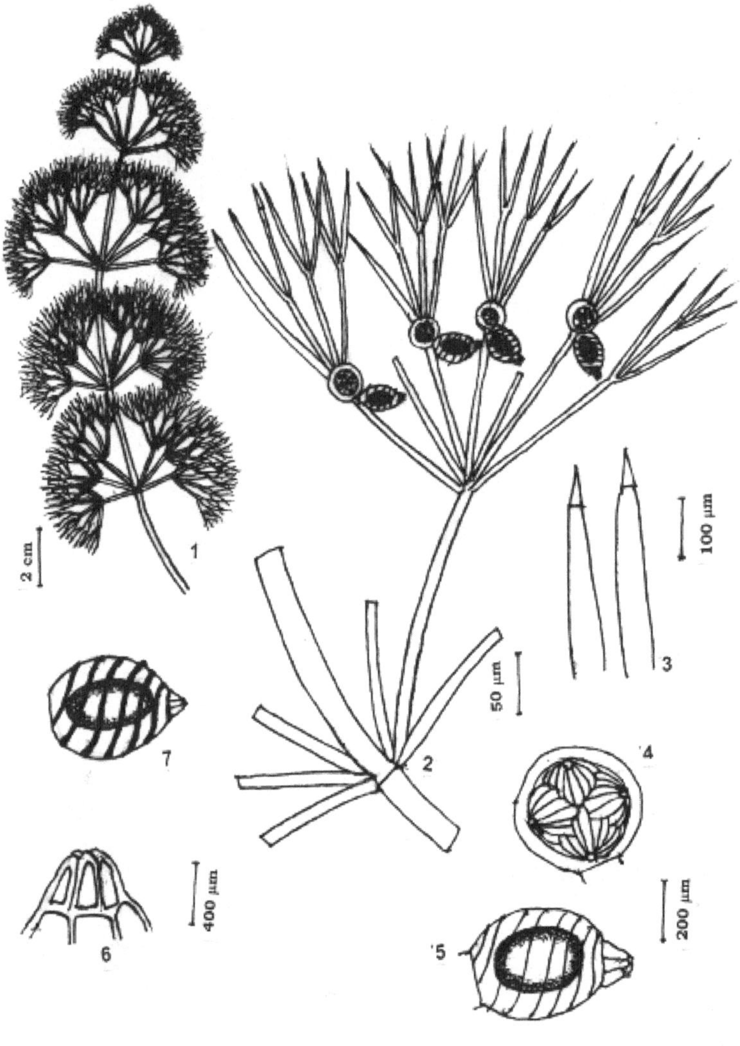

Plate 35a *N.tenuissima*. Fig. 1 = A plant; 2 = A whorl with a branchlet enlarged; 3 = Dactyls; 4 = Antheridium; 5 = Oogonium; 6 = Corona cells; 7 = Oospore.

5. PLANTS STUDIED FROM PACHAMALAI HILLS, SRI PERUMPUDUR AND VRIDHACHALAM OF TAMILNADU

More interesting members of Charophyta have been observed from Pachamalai Hills of Tamil Nadu. Pachamalai Hills are present in Tiruchi District of Tamil Nadu and the maximum height is 3500 feet from MSL.

The plants observed from Pachamalai: Only one species of *Nitella* and three species of *Chara* have been observed during July 2000 near Peria Pankalam, a small village, in paddy fields. During October 2000, a few species of Charophyta have been observed in these areas. In general, the plants are small and bushy. Herbarium specimens and liquid preservations are deposited at 277, I South Cross Road, Mariappa Nagar, Annamalai Nagar – 608 002, Tamil Nadu. The observations of these species have been published in Glympses of Indian phycology edited by Keshri, J.P. and Kargupta, A.N. and printed by Bishen Singh Mahendra Pal Singh, Dehradun, India in 2005, pages 61 to 74.

1. ***Nitella charoides* D. Subramanian Glimpses of Indian Phycology, 2005, p. 6. (Plate 36 and Plate 40, Fig. 1).**

   Plants,6 to 12 cm in height, 8 branchlets; each furcates into 3 to 5 dactyls; dactyl 2 celled; tip cell short and conical; monoecious; either antheridium or oogonium at a node; oogonium yellowish orange, 450 to 500 μm in length and 300 to 350 μm in breadth; coronula cells short and blunt; 55-40 μm in length and 35 to 40 μm in breadth; 8 to 10 spiral cell convolutions; antheridium orange; 400 to 420 μm in diameter; octascutate.

   Isotype No. DS 265, dated 04.07.2000 deposited at277, Mariappa Nagar, Annamalai Nagar – 608 002, Tamil Nadu. HolotypeNo. 11a at B.S.I., CBE, T.N.

   Habitat: In paddy fields on moist soil with 1 or 2 cm water levels near Peria Pankalam of Pachamalai Hills collected on 04.07.2000 collection No. 265.

2. ***Chara zeylanica* var. *robusta* D. Subramanian, Glimpses of Indian Phycology, 2005, p. 64. (Plate - 37 and Plate 40, Fig. 2).**

Bushy dwarf and robust plant, 3 to 4 cm in height; dark green; branchlets recurved; main axis corticated, branchlets 10; each 6 segmented; 5 stipulodes at node of branchlets, longer than sex organs, unicellular and short, basal cell and 2 tip cells ecorticated and other cells corticated in a branchlet, sex organs at the 2 lowermost nodes of a branchlets. Oogonium globular and brown; small, 200 to 250 μm in length and 150 to 175 μm in breadth; 5 coronula cells; 7 to 8 spiral wall convolutions, antheridium orange, small; 160 to 175 μm in diameter; octascutate.

Isotype No. DS 266, dated 04.07.2000 deposited at 277, Mariappa Nagar, Annamalai Nagar – 608 002, Tamil Nadu.Holotype No. 11d. at B.S.I., CBE, T.N.

Habitat: In the paddy field wet soil with little water near Peria Pankalam of Pachamalai collected on 04.07.2000, collection No. 266.

3. ***Chara socotrensis* var. *incurvata* D. Subramanian, Glimpses of Indian Phycology, p. 64, 2005. (Plate 38 and Plate 40, Fig. 3).**

Medium sized plants, 8 to 10 cm in height, entire plant ecorticated; 9 branchlets, recurved, each 7 segmented, the lower and tip segments shorter, stipulodes short and 3 in number, slightly longer than sex organs, only at the lowermost 2 segments, sex organs at the lowermost 2 nodes only. Bisexual; oogonium orange yellow; medium sized, 350 to 450 μm in length and 300 to 330 μm in breadth; 10 coronula cells; short and blunt; 8 to 9 spiral wall convolutions; antheridia orange, globular, 300 to 350 μm in diameter; octascutate.

The plants resemble more or less to *C. socotrensis* but these plants are shorter and smaller in all characters.

Isotype : No. DS 267. Dated 04.07.2000 deposited at 277, Mariappa Nagar, Annamalai Nagar – 608 002, Tamil Nadu.Holotype No. 11c. at B.S.I., CBE, T.N.

Habitat: Collected in a paddy filed near Peria Pankalam of Pachamalai Hills on 04.07.2000. Collection No. 267.

4. ***Chara socotrensis* var. *ebracteata* D. Subramanian, Glimpses of Indian Phycology, 2005, p. 67. (Plate 39 and Plate 40, Fig. 2).**

Large plants when compared to the 2 species described; 10 to 12 cm in height; entire plant ecorticated; 9 branchlets at each node of main axis; branchlets slightly curved; 5 segmented; the tip segment smaller; sex organs only in the lower two nodes of the branchlets; stipulodes absent at all the nodes of branchlets; bisexual; sex organs smaller; oogonium yellowish green; 350 to 370 μm in length and 320 to 330 μm in breadth; 10 coronula cells; 8 to 9 spiral wall convolutions; antheridium orange; 300 to 340 μm in diameter, octascutate.

Isotype: No. DS 268, dated 04.07.2000 deposited in herbarium of author's home lab at Mariappa Nagar, Annamalai Nagar – 608 002, Tamil Nadu. Holotype No. 11b. at B.S.I., CBE, T.N.

Habitat: Collected at Pachamalai in a paddy field on 04.07.2000. Collection No. 268.

Discussion

Nitella charoides sp. nov. studied, resembles *N. gracilis* var. *confervasia f. conformis* (Nordst.) RDW in having only one furcation of the branchlets into 3 to 5 dactyls, 2 celled, and the tip cell of dactyl shorter and conical. But the species studied now is different from the former species in having internodes of axis as long as branchlets; primary rays ½ of the length of entire branchlet; absence of fertile whorls and shorter and bushy appearance. The species studied here is also differing from *N. gracilis* subsp. *gloeostachys* var. *asagrayana* (Schaffn. ex. Nordst.) RDW in bushy habit, in having 8 branchlets at the nodes of main axis, primary rays ½ of the total lengths of branchlets and in the absence of special fertile whorls.

The specimen studied from Pachamalai Hills is not resembling in many characters with any of the species of *Nitella* described by Wood and Imahori

(1965). Therefore, the species studied is new to science and one of the simplest forms of *Nitella* so far studied from various parts of the world and as it is resembling an ecorticated *Chara*, for external appearance it is named as *N. charoides* sp. nov. The second specimen of *Chara* resembles *Chara zeylanica f. elegans* (A.Br. ex T.F.A.) H & J. Gr. (*C. palmeri* R.D.W.) in general appearance. But unlike this species the species of Pachamalai Hills is a dwarf plant with more branched and bushy habit with smaller branchlets which are equal to internodes of main axis in length. Instead of *C. zeylanica f. elegans* having sex organs only at the lower 3 nodes of 8 segments, the present specimen has sex organs only at the lower 2 nodes of 6 segments of the branchlets. Therefore, the specimen studied is considered as a new variety of *Chara zeylanica* as it is not at all resembling to any of the varieties of *Chara zeylanica* described by Wood and Imahori (1965) and it is named as *Chara zeylanica* var. *robusta* var. nov. D. Subramanian (Plate – 40).

The two ecorticated members of *Chara* resemble more or less to *C. socotrensis* Nordst. In *C. socotrensis* the plants are taller and the internodes of the axis are 2 times longer than the branchlets. The branchlets have only 3 segments. The sex organs are present in the lower 2 branchlets nodes (either at lower 1 or sometimes 2 nodes). In the third and fourth specimens, the sex organs are present in the lower 2 nodes of the branchlets. The third specimen has 7 segments and the fourth specimen 5 segments in a branchlet. In the third specimen, the branchlets are more recurved towards the axis and in the fourth specimen, the branchlets are slightly recurved towards the axis. The third specimen is smaller not only in general habit but also in all morphological and reproductive characters when compared to the fourth specimen. Therefore, the two specimens may be considered as two distinct varieties of *Chara socotrensis* and named as *Chara socotrensis* var. *incurvata* var. nov. (Plate 3) and *C. socotrensis* var. *ebracteata* var. nov. (Plate – 4) respectively.

A robust form of *Chara zeylanica* has been collected (Collection No. 264, dated 30.01.2000) from the fresh water ditch of Sriperumpudur near Madras (Plate 6). It is a diplostephanous form with prominent stipulodes, bracts and bracteoles and the cortex has hooked bristles all over the surface.

This species is comparable to the robust forms available in backwater canals of mangrove vegetation and salt water ditches of Rameswaram and other coastal areas of Tamil Nadu. But such a robust form is collected in fresh water ditch at Sriperumpudur and it is named *C. zeylanica* var. *spinosa.* D. Subramanian, 2005. Glimpses of Indian Phycology, p. 71 (Plate- 40).

In another isolated large fresh water ditch among the uncultivated waste lands near Sriperumpudur, a large number of *Nitella* plants were collected. They were identified as *Nitellafurcata* subsp. *flagellifera f. patula* (J. Gr. & G.O.A.) R.D.W. and *N. furcata* subsp. *flagellifera* (J. Gr. & G.O.A.) R.D.W. (Collection Nos. 268 to 275; dated 30.01.2000). Herbarium specimens are deposted at author's laboratory.

In Virudhachalam Taluk of Tamil Nadu, three different forms of *Chara corallina* (Collection No. 260 to 263, dated 25.01.2000) and a form of *C. braunii* have been collected in a fresh water lake of Kappankulam village during the month of June 2000 when all the other lakes and tanks completely dried up. Therefore, these forms of *Chara corallina* and *C. braunii* were dwarf and poorly available (Plate - 41).

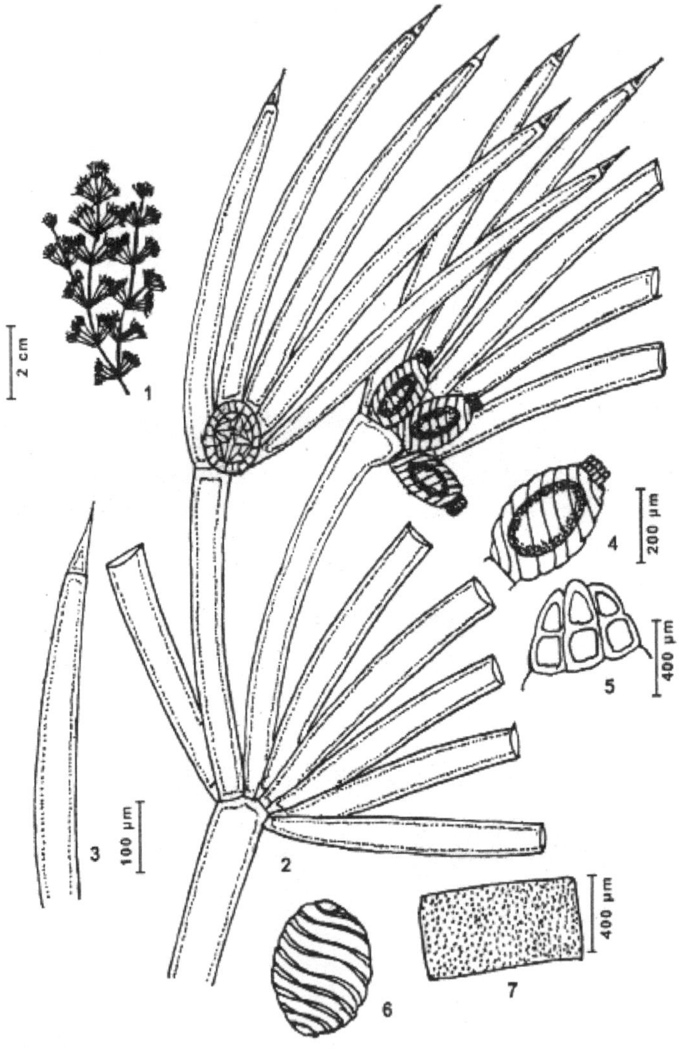

Plate 36 *Nitella charoides* sp. nov.; Figs. 1-7. 1: Habit; 2: Branchlets with sex organs (enlarged); 3: A dactyl; 4: Oogonium; 5: Coronula cells; 6: Oospore; 7: Granular oospore membrane.

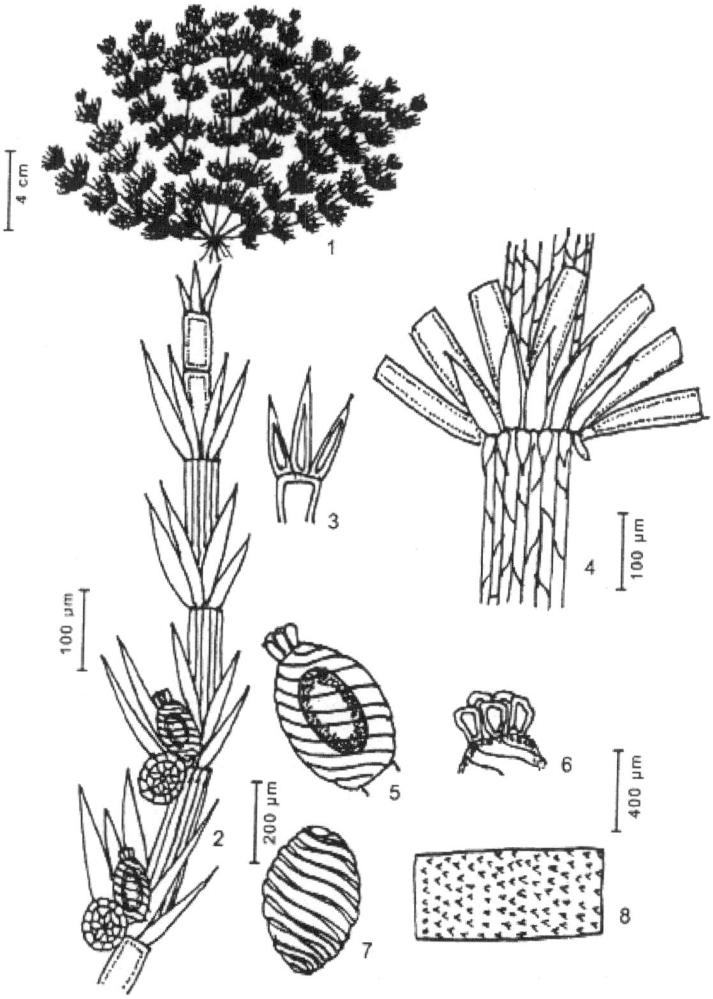

Plate 37 *Chara zeylanica* var. *robusta* var.nov.; Figs. 1-8: 1: Habit; 2: A Branchlet with sex organs; 3: Tip cell of branchlet; 4: Main axis with 2 rows of unequal stipulodes; 5: Oogonium; 6: Coronula cells; 7: Oospore; 8: Slightly papilloseospore membrane.

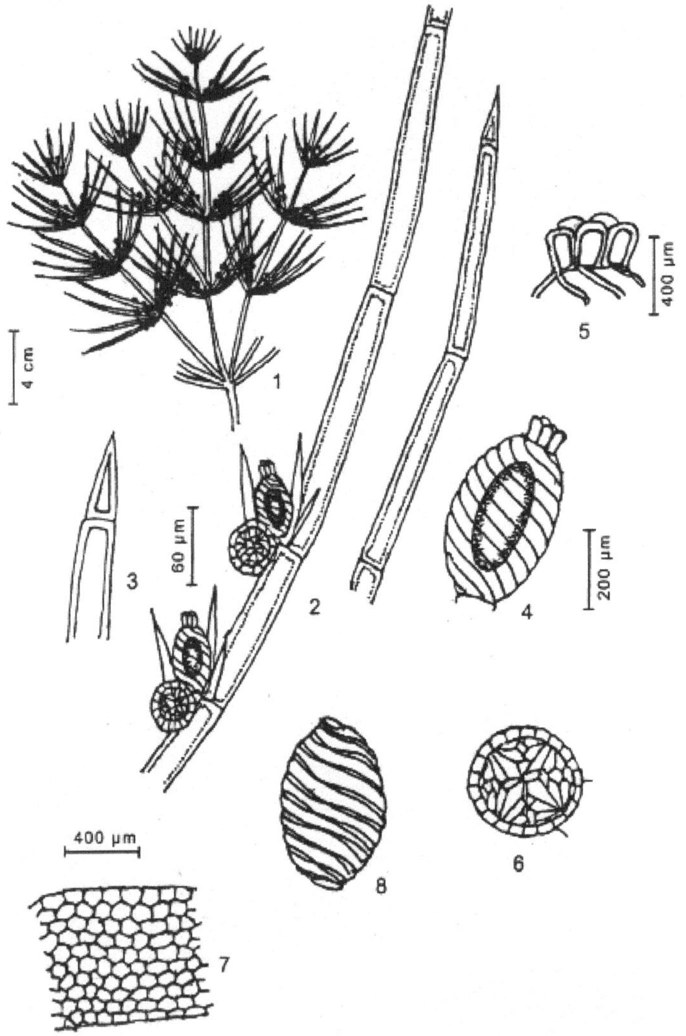

Plate 38 *Chara socotrensis* var. *incurvata* var. nov., Figs. 1-8 1: Habit, 2: A branchlet with sex organs enlarged; 3: Branchlet tip cells; 4: Oogonium; 5: Coronula cells; 6: Antheridium; 7: Netted oospore membrane; 8: Oospore.

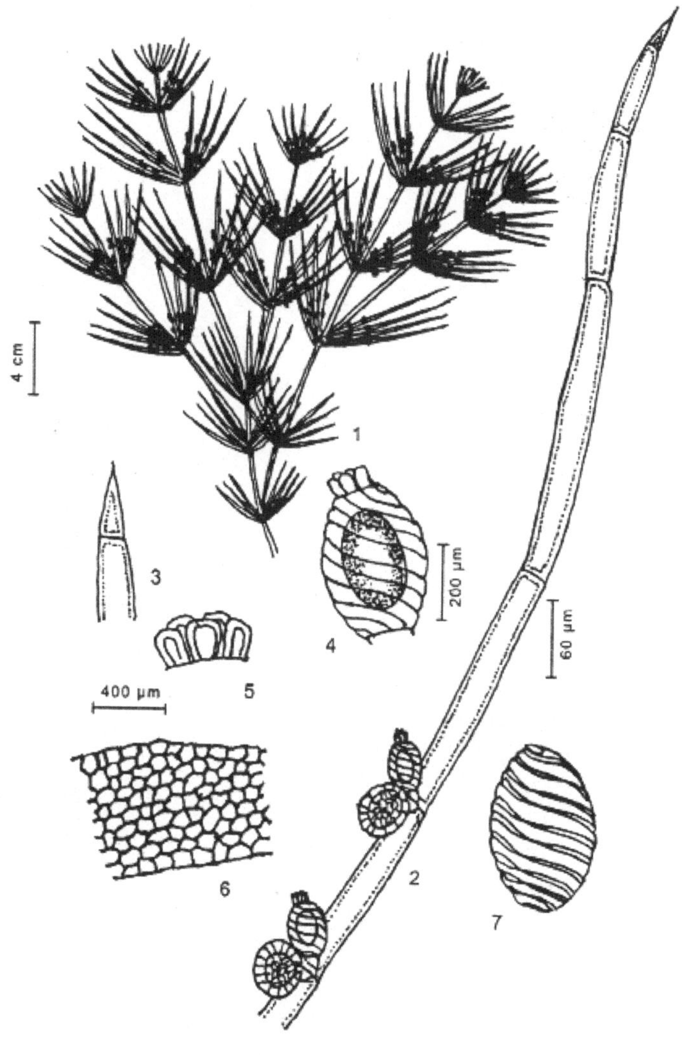

Plate 39 *Chara socotrensis* var. *ebracteata* var. nov., Figs. 1-7: 1: Habit; 2: A branchlet with sex organs; 3: Branchlet tip cells; 4: Oogonium; 5: Coronula cells; 6: Netted oospore membrane; 7: Oospore.

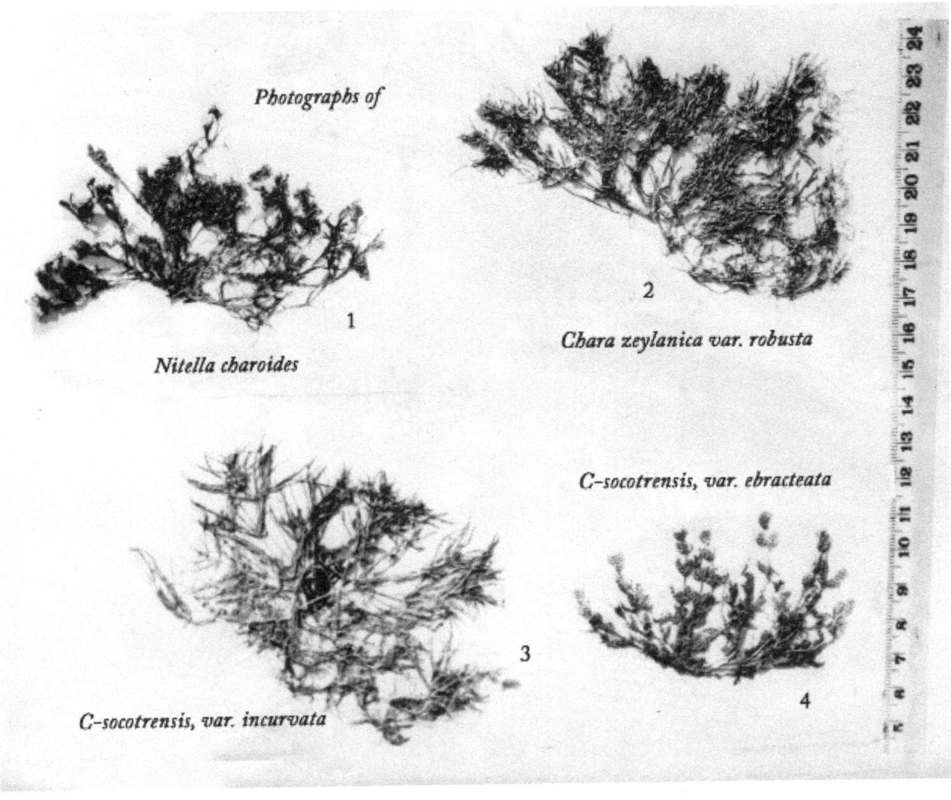
Photographs of
1
Nitella charoides
2
Chara zeylanica var. robusta
3
C-socotrensis, var. incurvata
C-socotrensis, var. ebracteata
4

Plate 40

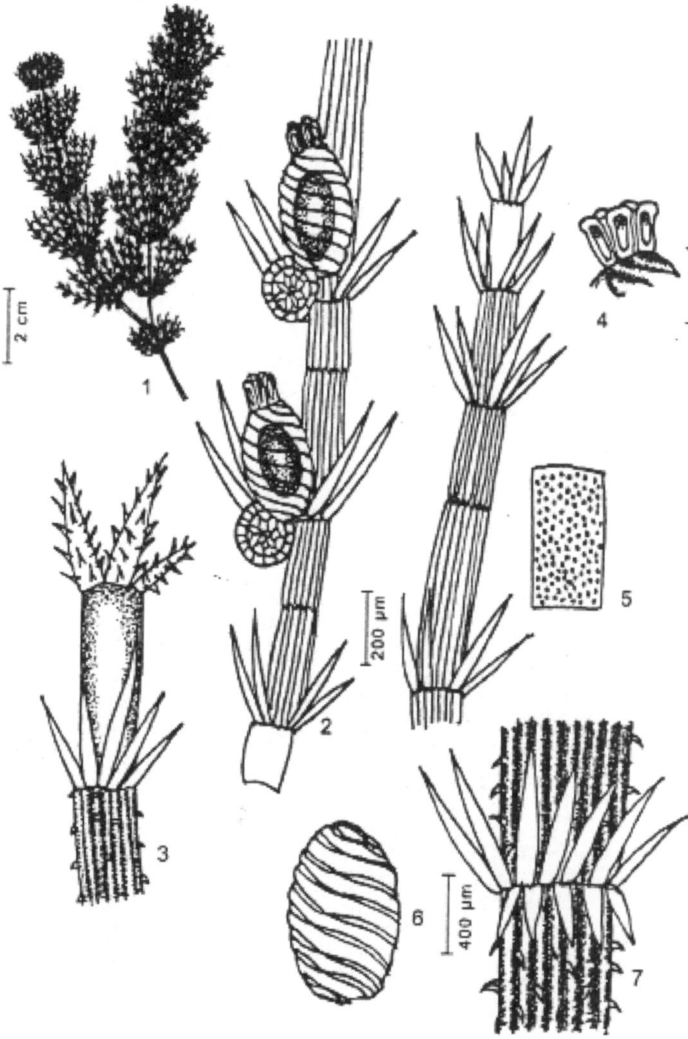

Plate 41 A robust form of *Chara zeylanica* Figs. 1-7: 1: Habit; 2: A branchlet with sex organs; 3: Tip cells of branchlet; 4: Coronula cells; 5: Dotted oospore membrane; 6: Oospore; 7: Main axis with spines and stipulodes.

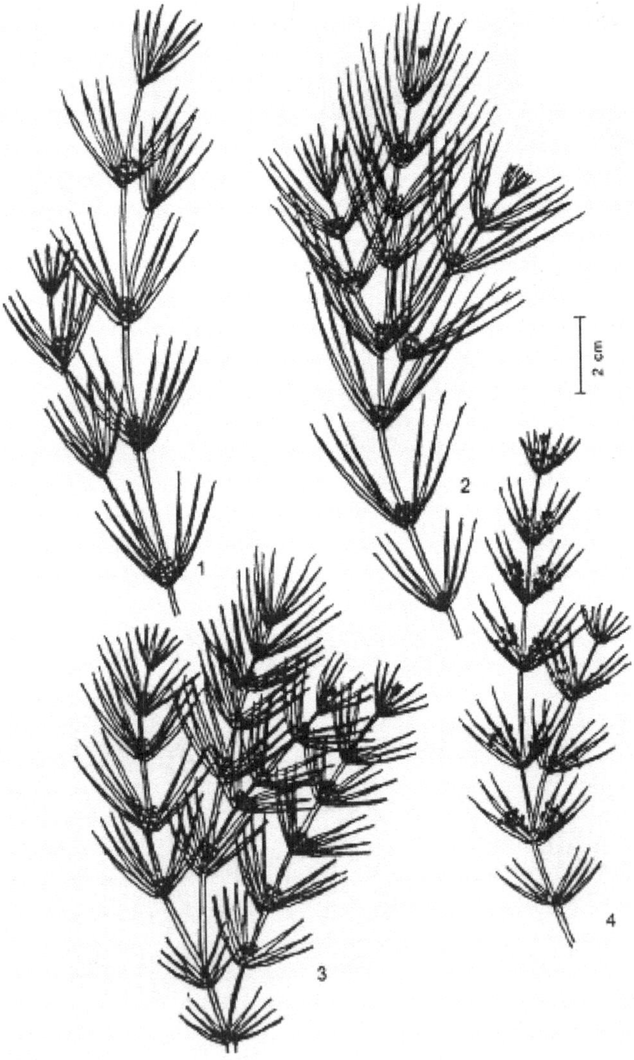

Plate 42 Fig. 1-3: Three different forms of *chara corallina*; 4: *Chara braunii.*

6. REVISION OF TAXONOMY OF TOLYPELLA FROM TAMIL NADU

In the author's previous book "Monograph on Indian Charophyta," 3 members of *Tolypella* have been described, namely *Tolypella nidifica* var. *glomerata* (Desv.) R.D.W. (Plate – 25), *Tolypella intricata f. prolifera* (Ziz ex. A.Br.) R.D.W. (Plate – 26) and *Tolypella intricata* var. *dispersa*. D. Subramanian (Plate 27) as occurring in India.

Generally, there are separate sterile whorls and fertile whorls in *T. nidifica* and *T. intricata*. The sex organs are present only on fertile whorls. But, in *I. intricata* var. *dispersa* collected and described for the first time in South India, particularly in Tamilnadu by the author, there are no separate fertile and sterile whorls. All the branchlet whorls are similar and the sex organs are present on the upper part of the branchlet whorls. The author has drawn a peculiar oospore wall ornamentations in this species. Because of the absence of distinct fertile whorls and presence of sex organs in the upper whorls of branchlets, this plant may not be included either under *Tolypella nidifica* or *Tolypella intricata*. It may be treated as a separate species and the author renamed this plant as *Tolypella dispersa*. D. Subramanian (Plate - 43). Now, the author is giving more importance because of the taxonomic significance of this plant and further attempts will be undertaken to thoroughly search this area, Mavidanthal and its neighbourhood and Vridhachalam, Chidambaram and Cuddalore of Cuddalore District as a whole to find some more plants of this species.

The habits of the plants had been illustrated in 3 *Tolypella* like species of *Nitella*, namely *N. cristata*, *N. Imahori* and *N. strutthioptila* and 18 members of *Tolypella* coming under *T. nidifica* and *T. intricata* in their "Iconograph of charophyta," by wood and Imahori in 1964. These plants show separate sterile and fertile whorls and sex organs are present only on fertile whorls.

One bitter experience the author is facing is the very weak and delicate thallus of *T. dispersa*. Neither Herbarium preparations nor liquid preservations are possible and these specimens become collapsed within a short time. When the author is successful in a long standing herbarium preparation of so many other taxa of Charophyta, in a few cases, it is not possible. Therefore, further and further, we have to collect them and make the Herbarium. Therefore, wherever possible, it is better to take natural or microscopic photograph of them.

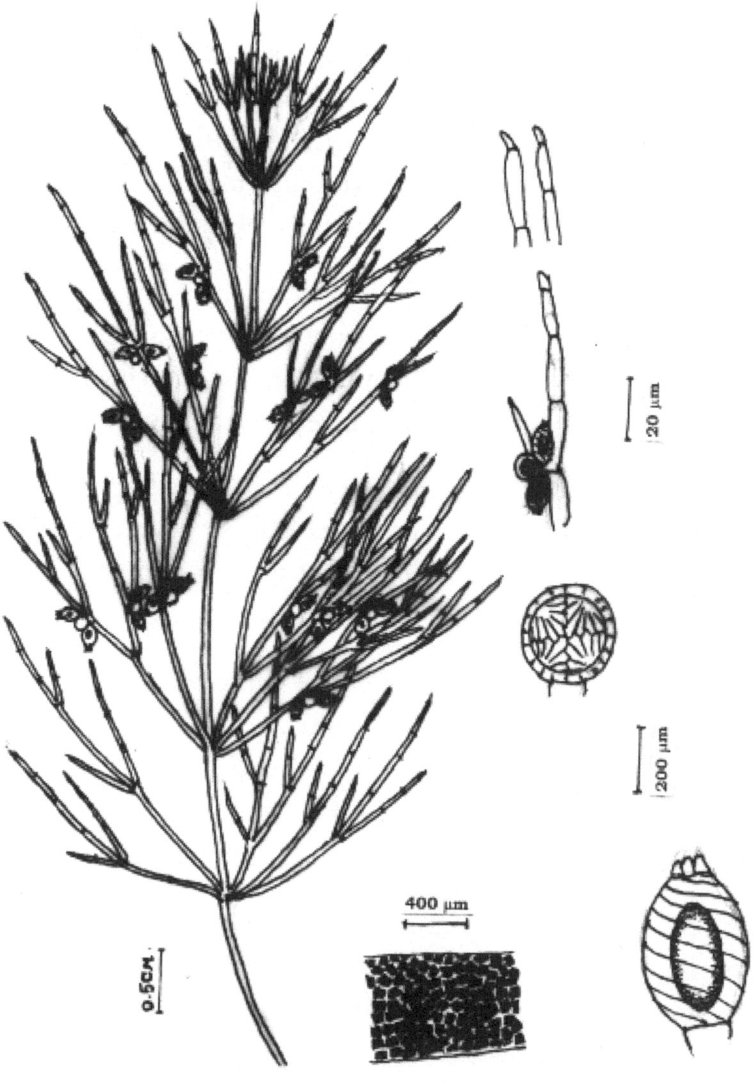

Plate 43 *Tolypella dispersa*. Fig. 1 = A plant enlarged with sex organs; 2 = Dactyls; 3 = sex organs; 4 = Antheridium; 5 = Oogonium; 6 = Oospore membrane wall ornamentation.

7. STUDIES ON CHAROPHYTES FROM ANDHRA PRADESH

Except some fragmentary reports on Charophyta from Andhra Pradesh, the Hyderabad division and neighbouring areas still remain taxonomically unexplored. The works of M.O.P. Iyengar, Sundaralingam and others on South Indian Charophytes were mostly extended to Tamilnadu and rarely to Karnataka, Andhra Pradesh and Kerala (Pal, et al., 1962; Sundaralingam, 1959 and Subramanian, 1983).

The author has attended an Advanced level Institute on recent trends in plant Genetics at OsmaniaUniversity, Genetics Department, Hyderabad from 5 to 25 September 1983. During holidays and spare times, he searched for charophytes at various places of Osmania University Campus.

Plants grow luxuriously even in small ditches and experimental cement rectangular pots (3 x 1½ feet size, and 1¼ to 1½ feet height) filled with water. In the soil at the bottom of the pots, *Chara* species germinated and grew as green beds.

Holotypes and herbarium specimens are deposited at author's laboratory, Mariappa Nagar, Annamalai Nagar – 608 002, Tamil Nadu.

The plants are briefly described below:

1. *Chara vulgaris f. glabra*. forma. nov. D. Subramanian (Plate - 44).

Plants 18 to 20 cm tall, erect, rarely branched, monoecious; stem 660 to 670 μm thick; main axis corticated, whorls of 6 branchlets, each branchlet of 4 to 5 segments and ecorticated, bracts and bracteoles absent; stipulodes 12, rudimentary, sex organs orange, at the first nodes of branchlets, Oogonium 330 to 350 μm in length and 250 to 260 μm in breadth, spiral cells with 8 convolutions,antheridia 250 to 260 μm in diameter.

Collected by the author in cement pots with fresh water at OsmaniaUniversity campus in an experimental research garden. Collection Nos. 183 to 186 dated 08.09.1983. Holotype deposited at author's laboratory in his house. Isotype No. 80 at B.S.I., CBE, T.N.

This plant is a medium sized light green coloured and apparently, glabrous form. In the presence of corticated axis and ecorticated branchlets with fewer number of segments, it resembles *Chara vulgaris* var. *inconnexa f. hippelliana* (Villi.) R.D.W. (*C. contraria* subsp. *hippelliana*, Villi.), but in larger plant size, it differs from the above plant with more number of branchlet segments and presence of sex organs only on the first nodes of branchlets, and so named *C. vulgaris, f. glabraforma*. nov.

2. *Chara vulgaris, f. simplex forma*. nov. D. Subramanian (Plate-45)

Plants small and bushy with large number of branches arising at the base of the plant. Monoecious; stem 330 µm thick, corticated, whorls of 6 branchlets each; of 6 segmented and ecorticate, stipulodes 6, bracts and bracteoles absent; Oogonium orange and smaller, 250 to 260 µm in length and 150 to 160 µm in breadth, spiral cells with 8 convolutions; Antheridia orange, 200 to 210 µm in diameter; sex organs at the basal nodes of branchlets.

Collected at cement pots by the author on 08.09.1983. Collection Nos. 187 to 194. Isotype No. 81 at B.S.I., CBE, T.N.

This species resembles in cortex of axis and branchlets nature to *C. vulgaris* var. *inconnexa f. hippelliana* (Vihl.) R.D.W. but in smaller and bushy appearance of plants, and 5 segmented branchlets, the present plant is differing. It is, therefore, named here as C. vulgaris f. simplex *forma.nov.* Voucher herbarium specimens are deposited at author's laboratory.

3. *Chara braunii f. delicatula forma*. nov. D. Subramanian (Plate-45)

Monoecious; medium sized delicate plants, rarely branched, whole plant ecorticated; stem 350 µm in thick, whorls of 6 branchlets, each branchlet 6 segmented, bracts and bracteoles at each node 5, 6 or 7; Stipulodes 6; Sex organs orange, Oogonium 300 to 310 µm in length and 250 to 260 µm in breadth; spiral cells with 9 convolutions; Antheridium 250 to 270 µm in diameter; sex organs at the basal 2 nodes of branchlets only.

Collected by the author in a large fresh water canal, by the side of OsmaniaUniversity on 15.09.1983. Collection Nos. 195 to 200. Herbarium specimens are deposited at author's house laboratory. Isotype No. 82 at B.S.I., CBE, T.N.

This plant resembles the type species *Chara braunii* Gm, in most of the characters and it differs from it in delicate and soft nature of the plants and coronae of mostly one and rarely 2 blunt cells only.

4. *Chara zeylanica f. robusta* forma. nov. D. Subramanian

Plants robust in texture, large, erect, 18 to 20 cm tall, rarely branched; monoecious; stem 840 μm thick; whorls of 8 branchlets, each branchlet of 9 segments, whole plant corticated excepting the basal segment of branchlets, stipulodes in 2 tiers, upper longer and lower shorter, spine cells longer, 600 to 800 μm long, bracts and bracteoles prominent; Oogonium large, 515 to 530 μm in length and 300 to 315 μm in breadth; spiral cells with 10 convolutions; Antheridium quadriscutate, 330 to 340 μm in diameter.

Collected by the auther in a fresh water ditch near Botanic Gardens, OsmaniaUniversity, Hyderabad on 15.09.1983. Collection Nos. 201 to 205. Herbarium specimens are deposited at author's house laboratory at Mariappanagar, Annamalai Nagar – 608 002, Tamil Nadu.

This plant is a peculiar form, as it has more number, longer and more robust nature of spine cells. Spine cells are more prominent than those forms of *C. zeylanica* particularly *f. inconstans* (Kutz) Groves and *f. hildebrandtiana* (A.Br.) Comb,described bySundaralingam (1959).

5. *Chara vulgaris* var. kovertinensis var. nov. D.Subramanian (Plate - 46)

The herbarium specimens of this plant collected at Kovertinagiri, Chittor District of Andhra Pradesh, are deposited at 277, Mariappa Nagar, Annamalai Nagar – 608 002, Tamil Nadu. Isotype No. 83 at B.S.I., CBE, T.N.

Plants medium sized, 12 to 15 cm tall, more branched, internodes 3 to 4 times longer than the branchlet whorls, a much relaxed beautiful plant; at each node of axis, 8 to 10 branchlets; main axis corticated, branchlet segments ecorticated, 4 segments in a branchlet, bracts longer and unequal, sex organs only at the basal two branchlet nodes.

This plant resembles *Chara vulgaris* var. *kirghisorum*, f. *hungaria* (Fil.) R.D.W. described by Wood and Imahori (1965) (Icone – 19) to some extent, but differs in the narrow and elongated segments of branchlets, larger size of branchlet whorls and the much reduced internodal lengths in comparison to whorls of branchlets.

6. *Nitella hyalina*(DC) Ag

A form of *Nitella hyalina* was collected from Bangarupalem of Chittor District of Andhra Pradesh on 06.01.1992 and the herbarium specimens of this species are deposited at DDE Wing, Botany Department, AnnamalaiUniversity but the various parts of this plant are not clear.

7. *Chara tomentosa*. L. em. var. *gigantea* (Plate - 47) var. nov. D. Subramanian

The herbarium specimen of this species collected from Jongolu, Krishna District of Andhra Pradesh in 1995 has been deposited at DDE Wing, Botany Department, AnnamalaiUniversity.

Plants tall, rarely branched, 25 to 28 cm high, spiny in nature. Internodes 2 times the length of branchlet whorls, 10 branchlets at each axis node, 9 segments in each branchlet, axis corticated with vertically pointed spine cells; bracts and bracteoles are small, stipulodes dotted in 2 tiers. Sex organs at the basal three nodes of branchlets, at each node one antheridium below and one oogonium above, antheridium light green, octascutate, oogonium larger, 10 spiral wall convolutions, light greenish orange.

This is a beautiful plant, suitable to be grown in glass jars, because of its magnificent plant form. This is named herein as a new variety, var. *gigantea*.

Wood and Imahori (1965) have drawn a few parts of this plant along with general habit of the plant but no complete descriptions are provided for this species (Icone 70 of Iconograph (Vol. II) of Wood and Imahori, 1965).

In this plant no spine cells and stipulodes are clearly seen and this plant is monoecious. The segments of branchlets and main axis are corticated. The two tip segments alone are ecorticated. In the present plant also, almost all the characters are agreeing with those of *Chara tomentosa*. Above all, the general appearance of the present plant is almost resembling that of *C. tomentosa* given by Wood and Imahori in their book. The spine cells and bracts and bracteoles give this plant a tomentose appearance.

The observation of this species is a first record for India.

8. *Chara setosa* Klein ex. Willd.

A medium sized plant was collected in this species from MusiRiver, R. Nagar, Hyderabad, A.P. Another medium sized form of this species was collected from Chittoor District.

9. *Chara zeylanica* Klein ex. Willd.

A form of this species was collected from Chenmur, Adilabad District, A.P.

The above said two forms of *C. setosa* and one form of *C. zeylanica* are deposited as herbarium specimens in DDE Wing, Botany Department, AnnamalaiUniversity, Tamil Nadu.

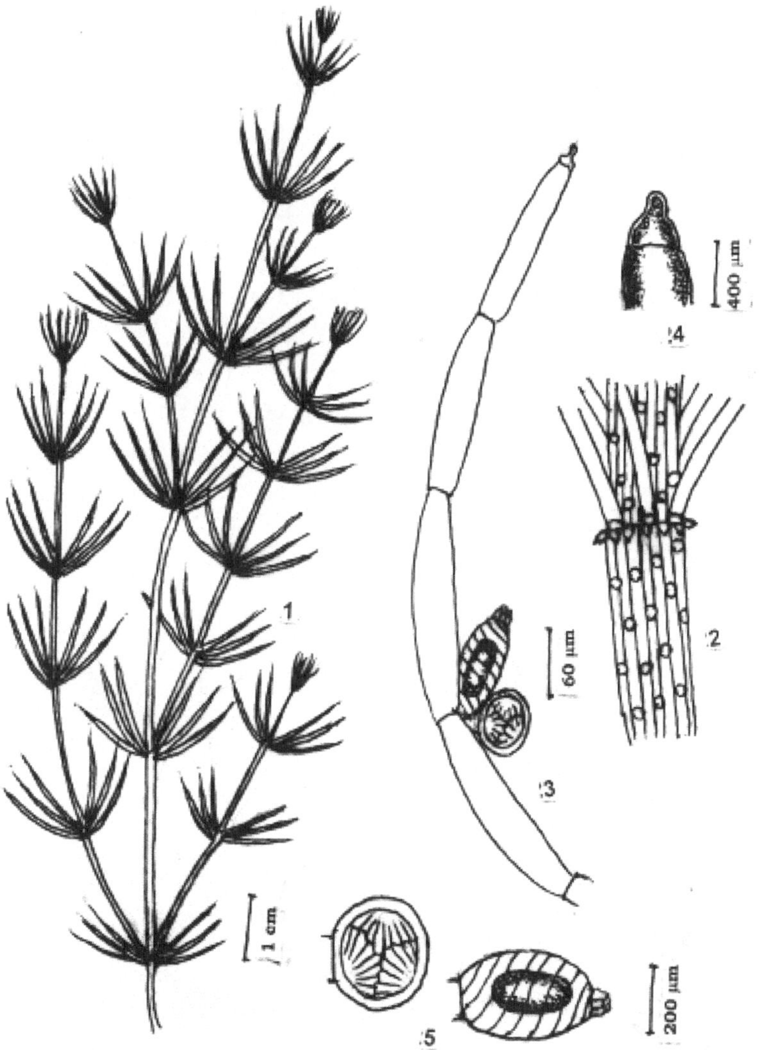

Plate 44 *Chara vulgaris. f. glabra*. Fig. 1 = Plant; 2 = Axial node; 3 = A branchlet; 4 = Branchlet tip; 5 = sex organs.

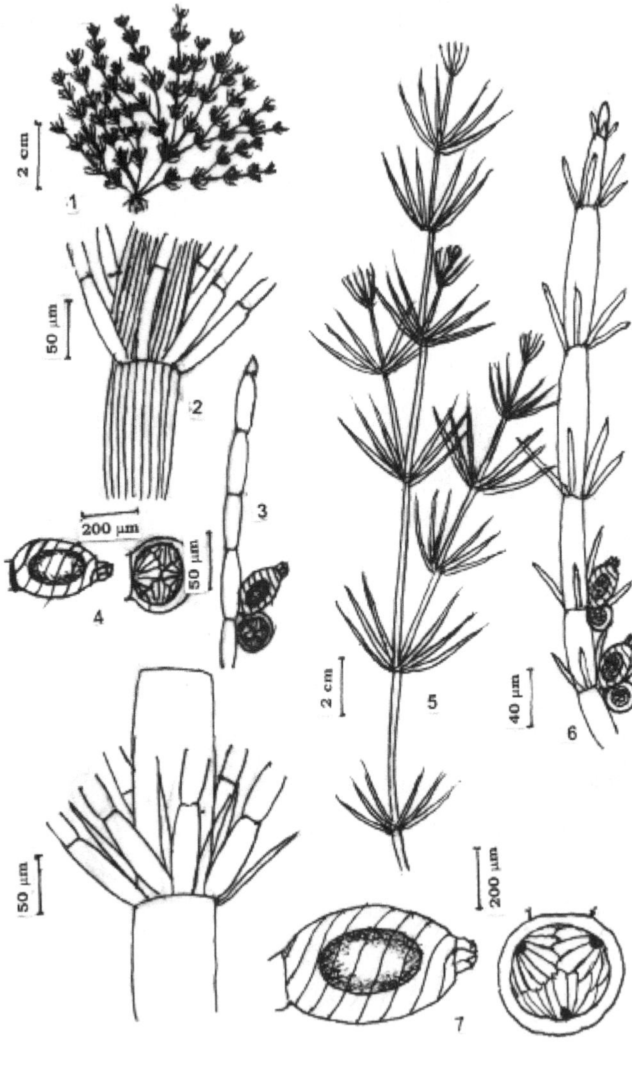

Plate 45 *C. vulgaris.f. simplex* Figs. 1 to 4 = *C. vulgaris. f. simplex*. Fig. 1 = A plant; 2 = Axial node; 3 = A branchlet; 4 = sex organs. Figs. 5 to 7 = *C.braunii. f. delicatula*. Fig. 5 = Plant; 6 = A branchlet; 7 = sex organs.

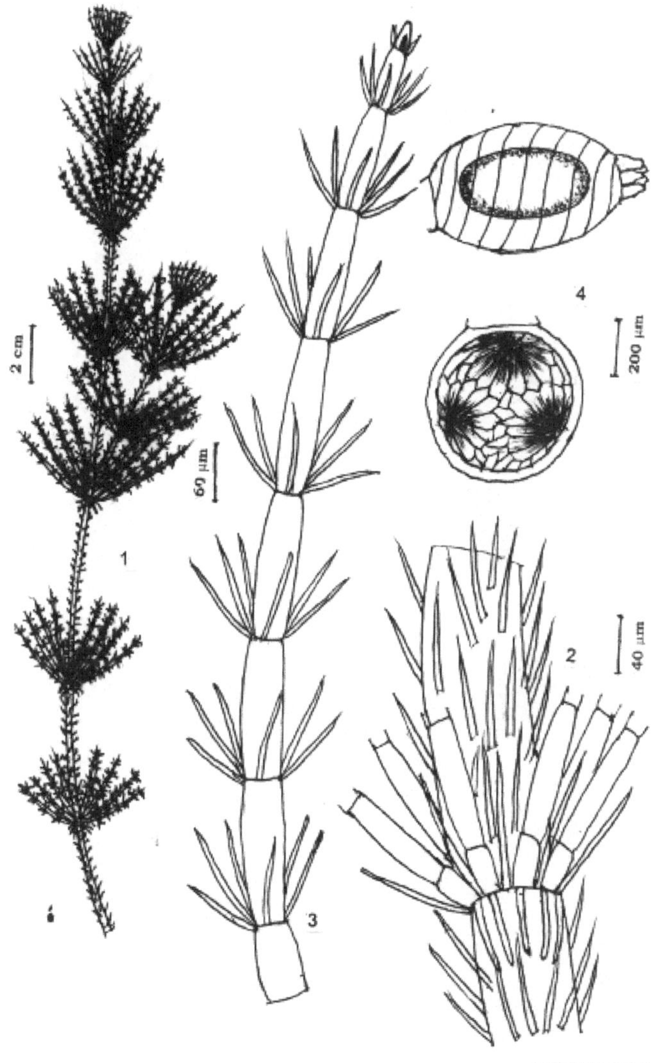

Plate 46 *Chara vulgaris. var. kovertinensis.* Fig. 1 plant; 2 = Axial node; 3 = A branchlet; 4 = sex organs

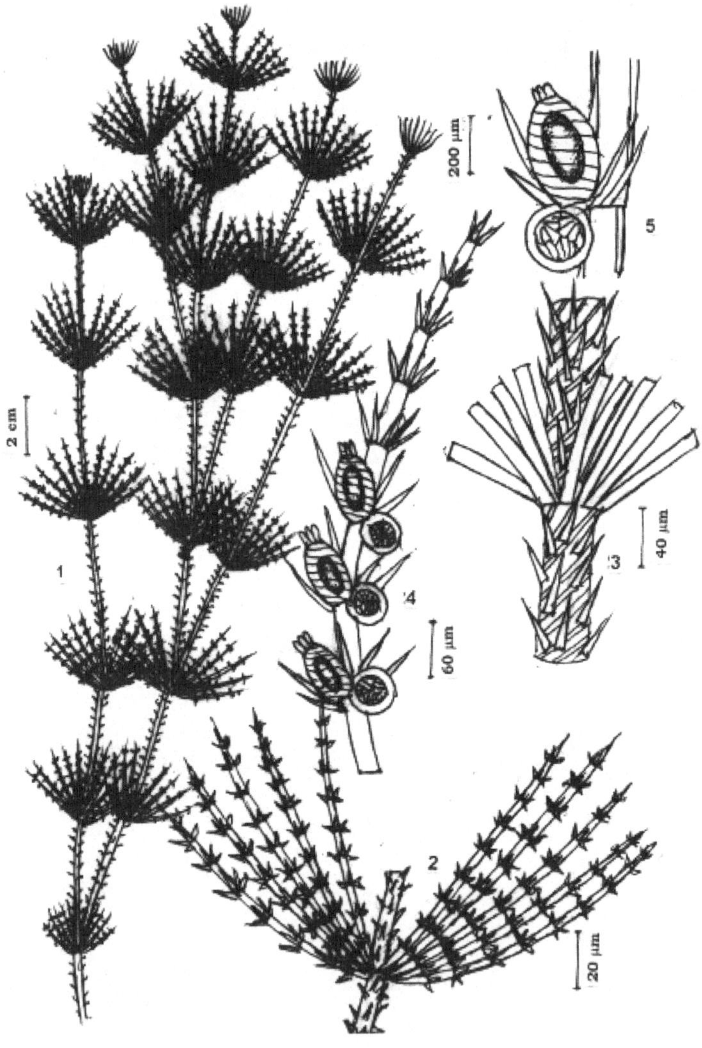

Plate 47 *Chara tomentosa*. var. gigantea. Fig. 1 = A plant; 2 = Axis with a whorl of branchlets; 3 = Axial node enlarged; 4 = A branchlet; 5 = sex organs.

8. STUDIES ON CHAROPHYTES FROM KERALA

1. ***Nitella substilissima var. umbellata*** **D. Subramanian in National Seminar on Climatic Change and Environmental Challenges, 2009. Botany Department, AnnamalaiUniversity, Tamil Nadu, Article 2.13, p. 33. (Plate - 48).**

This plant was collected from a large ditch near Thiruvananthapuram, Kerala. Holotype and Vaucher specimens are deposited at author's lab, Mariappa Nagar, Annamalai Nagar – 608 002, Tamil Nadu.

Plants delicate, more branched 10 to 12 cm tall, main axis slender and thin, 300 μm in breadth; internodes 2 to 4 times longer than the whorls of branchlets; branchlets of only one type, 6 per node, primary ray more than ½ of the entire branchlet length, primary ray forking into 4 to 5 dactyls and one secondary ray which again forking into 4 dactyls; Dactyls almost equal in size; 2 celled, end cell small and mucronate; sex organs not seen.

This plant resembles *Nitella substilissima* A. Br. em, in general appearance but the nature of whorls and number of secondary rays and dactyls are different in the present plant. This plant studied here is larger, more branched and sterile. Now, it is treated as a new variety and named herein *N. substilissima var.umbellata* var. nov.

2. ***Chara corallina*** **Klein ex. Willd, em. (Plate 16)**

The herbarium specimens of this plant, collected from Sathupally, Kammam District of Kerala in a small fresh water ditch, have been deposited at DDE Wing, Botany Department, AnnamalaiUniversity, Tamil Nadu.

Plants bushy, arising from a basal root stalk, 4 to 5 cm. tall, rarely branched; entire plant ecorticate, 5 branchlets at each node, each branchlet 5 segmented, the tip of branchlets each with 2 to 3 spine cells; No stipulodes, bracts and bracteoles; bisexual, sex organs at the base of the nodal region of axis.

This plant is dwarf, bushy and small unlike the species and varieties (*C. corallina*) so far studied from various parts of the world. This plant resembles *C. corallina. f. kyusyensis* (Imah) R.D.W. (Icone 113) in the absence of stipulodes, bracts and bracteoles but *C. corallina f. kyusyensis*has longer branchlet whorls, rudimentary stipulodes rarely and sex organs present in the axis node or branchlet nodes. But, the present plant is not at all comparable to any variety or form of Wood and Imahori (1965) or those published so far by the author.

3. *Chara zeylanica* Klein ex. Willd, em.

The herbarium specimens, of this plant collected from Kovalam of Kerala, have been deposited at DDE Wing, Botany Department, AnnamalaiUniversity, Tamil Nadu.

Plants medium sized, 10 to 12 cm tall, light green, branched intermittently; 6 branchlet at each whorl, 6 segmented, basal segment shorter and ecorticate, all upper segments of branchlets and main axis corticated; sex organs at the second and third node of branchlets; oogonia are elongated barrel-shaped resembling to those of *Chara vandalurensis* Sundaralingam, stipulodes, bracts and bracteoles are also resembling to those of the above species. But in the present plant the basal segment is shorter and ecorticate and all the other segments corticated, as against to those of *C. vandalurensis* Sundaralingam.

The present plant is not so much related to any form of *C. zeylanica* var. *diaphana* described by Wood and Imahori (1965).

3. *Chara zeylanica* Klein ex Willd, em.

The herbarium specimens of the plants collected from Kovalam near Trivandrum of Kerala on 01.05.2008 have been deposited at DDE Wing, Botany Department, AnnamalaiUniversity, Tamil Nadu.

Plants taller, branched 20 to 27 cm high, internodes 2 times the length of the branchlet whorls; 10 to 11 branchlet in each whorl, at each node of branchlets 4 to 5 short bracts and bracteoles, each branchlet 7 segmented, basal and upper 1 or 2 segments ecorticate, basal segment short, all other segments and main axis corticated; stipulodes in 2 tiers, upper longer and lower shorter; sex organs at the 2nd and 3rd nodes of branchlets, one antheridium and one oogonium at each node; young oogonium slightly elongated globular but mature oogonium globular; Antheridium quadriscutate, 10 to 11 convolutions on oogonium wall.

This plant resembles to *Chara zeylanica f. filicaulis* (Roxb.) R.D.W. in general appearance of plants; but in a large number of characters it is different from it.

The author has not travelled through different parts of Kerala recently. Therefore, only a few plants have been studied from this state.

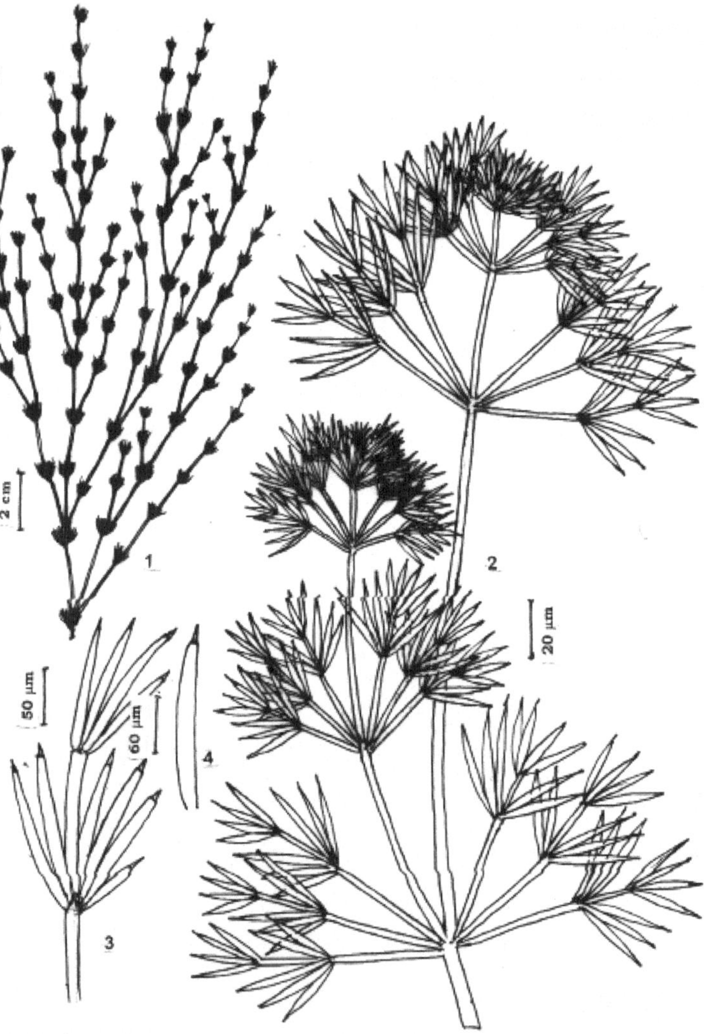

Plate 48 *Nitella substilissima var. umbellata* Fig. 1 = A plant; 2 = part of plant enlarged; 3 = A branchlet; 4 = A dactyl.

9. CHAROPHYTES FROM KARNATAKA STATE

There are 2 important cities, Bangalore and Mysore in this state. The climate is pleasant and salubrious throughout the year. Most of the fruit and vegetable yielding plants, ornamental plants and those usually growing in Western Ghats, particularly in Ooty and Kodaikanal, are cultivated in Bangalore and its neighbouring places. The soil is mostly red or calciferous. But, so far, there has been no intensive study of Charophytes in this state. Bharati and Chennaveeraiah (1983) have collected and studied the chromosomes of some forms of *Chara zeylanica* and *Lychnothamnus barbatus*. var. *gigantei* from this state.

Charophytes are not so much available in the city areas due to impure water and human and animal disturbances. Therefore, we have to go to remote places, particularly urban areas and hill sides and abandoned areas, where there are small lakes, ditches and water canals.

Some of the species of this group of plants were studied by the author from the Herbarium specimens kept at DDE, Botany Wing, AnnamalaiUniversity. Mr. Sheshadri Ramaswamy, a Nature Lover from Bangalore, has collected and sent the authorsome members of Charophyta from remote areas of Bangalore and Western Ghats of Karnataka state.

The plants, described below have been collected and handed over to the author by Mr. Sheshadri, R. from Bangalore and its neighbourhood in the first week of April 2015 and the author studied the specimens on 10.04.2015. There are 2 commonly available species, *Chara corallina* and *Nitella furcata*. There are 3 members, rare and more interesting, described below:

1. *Chara bauerii* A. Br. var. 1 (Plate - 49)

Plant light green, medium sized, rarely branched, 20 to 22 cm tall, main axis corticated, at each axial node 10 branchlets, eachbranchlet 5 segmented, stipulodes as many as branchlets; Bracts and bracteoles present, spinous projections are present on the main axis; sex organs at the basal 3 nodes, oogonium light green with 11 spiral wall convolutions, antheridium green;

branchlets ecorticated.Isotype at author's laboratory and Holotype No. 58 at B.S.I., CBE, T.N.

2. *Chara bauerii* A.Br. var. 2 (Plate - 50)

This plant is smaller than the previous plant, 15 to 17 cm tall; rarely branched, main axis twisted and corticated; stipulodes as many as branchlets of whorls, 12 branchlets at each axial node, each branchlet 4 segmented, all ecorticated, bracts and bracteoles present; sex organs at the two lower nodes of branchlets, oogonium orange, antheridium greenish yellow;2 plants of this species had been described in their book by Wood and Imahori (1964), namely 1) *Chara bauerii* A.Br. em. and *C. bauerii. f.muelleri* (A.Br.) R.D.W. but, the former plant alone described completely. The present species resembles those of Wood and Imahori.Isotype at author's laboratory. Holotype No. 59 at B.S.I., CBE, T.N.

3. *Nitella bastinii* T.F.A. (Plate - 53)

This is a peculiar plant, 7 to 9 cm tall. Sterile branchlet 5 at each axial node; sterile branchlet once furcate without sex organs, dactyl single celled; fertile whorls arise in the axil of sterile whorls, two or sometimes one, if 2, one above the other, smaller and globular, fertile whorl 2 times furcate finally into one celled dactyl; only oogonia seen each in first and second node of fertile branchlets; no antheridium seen; Therefore, the plant may be dioecious; we are trying to get the male plant.

This plant resembles to some extent *N. bastinii* T.F.A. described by Wood and Imahori (1964) in their monograph on charophyta, only by separate fertile whorls one above the other and dactyl one celled but in the nature of furcations of fertile whorls, the present plant differs completely from that of Wood and Imahori.

The occurrence of this species is a first record for India and very rare.

The plants described below have been collected and sent to the author by Mr. Sheshadri Ramaswamy from neighbouring places of Bangalore, Karnataka State from 2 to 07.05.2015. These plants have been examined, described and parts of the plants illustrated by the author. Herbarium specimens have been prepared at once and deposited at author's house laboratory, 277, I South Cross Road, Mariappa Nagar, Annamalai Nagar – 608 002, Tamil Nadu. Isotypes are deposited at Regional Botanical Survey of India, Southern Circle, Coimbatore, Tamilnadu. A few specimens are deposited at specimen bottles in 4% Formalin.Holotype No. 60 at B.S.I., CBE, T.N.

Plants collected at various places of KarnatakaState, like Tumkur, Nittur, Gubbi, Tiptur and Arsikere.

4. *Chara bauerii* A.Br. em (Plate - 51)

Plants 12 to 15 cm tall, branched, erect, thick or medium green, whorls of branchlets shorter than internode; main axis corticated, 8 to 9 branchlets at a node, branchlets ecorticated, 4 to 5 segmented, stipulodes as many as branchlets at a node, bracts and bracteoles medium sized and unequal; sex organs at the basal 2 nodes of branchlets; antheridium light green, oogonium orange, 8 to 9 spiral wall convolutions.

Already 2 varieties of *Chara bauerii* A.Br. have been studied from Bangalore and its neighbourhood along with *Nitella bastinii* T.F.A., a very rare species. These varieties were collected by Mr. Sheshadri, R. in the first week of April 2015.

5. *Chara leptopitys* A.Br. (Plate - 52)

Many erect, unbranched plants arising at the base of the plants and forming a thick colony; 6 to 8 cm tall, light green while alive and almost whitish grey when dry, 8 branchlets at the axial node; 6 segmented, all ecorticated, axis corticated; no stipulodes, bracts and bracteoles. Sex organs both antheridium and oogonium forming a cluster at the axial node and 1 antheridium and 1 oogonium at each of basal 2 nodes of branchlets; antheridium light green and oogonium dark brown to black.

There are 2 forms collected together, one taller and relaxed and the other shorter and bushy.

This species resembles *Chara leptopitys* A.Br. *f. subebracteata* (Nordst) R.D.W. in having sex organs at the axial nodes and branchlet nodes and also *C. corallina* Klein ex. Willd; em. But, in the presence of corticated main axis, this species resembles *C. leptopitys. Chara leptopitys* A.Br. *f.subebracteata* (Nordst.) R.D.W. is a dioecious member, whereas the present plant is monoecious. In the former plant, cluster of oogonia or antheridia more than 1 or 2 may be present at the basal nodes of branchlets, but in the present plant one antheridium and one oogonium is present at a particular node of branchlet. Stipulodes, bracts and bracteoles present in the former species but they are absent in the present plant. In the former species, spine cells are present on the axis, but they are absent in the present species. Therefore, the present plant may be a form of *C. leptopitys*, A.Br.

The study of this species is a first record for India. In Wood and Imahori's book (1964), the above mentioned plant alone, *C. leptopitys* A.Br. *f.subebracteata* has been described.Isotype at author's laboratory, Holotype No. 56 at B.S.I., CBE, T.N.

6. *Nitella duthieae* J. Gr. & Stephen (Plate - 54)

Plants 12 to 14 cm tall, branched intermittently, externally appearing a form of *N. furcata* Ag.em.

Thallus crowded and more furcate, dark green; at each node 6 branchlets arising, primary ray longer, more than ½ of the total length of branchlet, each forking into 5 to 6 secondary rays, 2 to 3 of them ending into 2 or 3 dactyls, dactyl short and conical, 2 or 3 of the secondary rays again forking to 3 tertiary rays, ending into dactyls, sex organs rare, one antheridium and one oogonium at the basal node of branchlets, both of them light green.

In the presence of forking patterns of primary rays and secondary rays, this present plant resembles *Nitella duthieae*, but like the type plant described by Wood and Imahori (1964), the thallus of the present plant are longer and more crowded resembling a large form of *N. furcata*. Besides, *N. duth-*

ieae has not been described completely but only various parts of this plant were illustrated by Wood and Imahori.

The occurrence of this species near Bangalore is a first record for South India.

Herbarium specimens are deposited at author's house laboratory at 277 I, South Cross Road, Mariappa Nagar, Annamalai Nagar – 608 002, Tamilnadu. Holotype No. 64, deposited at Botanical survey of India Southern circle at Coimbatore, Tamilnadu.

7. *Nitella hyalina* (D.C.) Ag.

Mr. Sheshadri has collected for the first time from neighbouring areas of Bangalore and Western Ghats border, 4 forms of this species. Even in the external appearance, the four forms may be clearly distinguished. It is clear that in KarnatakaState also like Tamil Nadu,there may be large number of forms, varieties and even related species of this group obtained, if thoroughly searched in remote areas. A few forms are resembling *Nitella congesta* (R.Br.) A.Br. and *N. plumosa* A.Br. Now, attempts are being made in hills areas of Western Ghats of Karnataka State. Likewise in various parts of Northern India floristic studies on Charophyta will be continued. As shown in the illustrations (Plates 30-32) the important differences of the forms of *N. hyalina* have been briefly described.

a. Form – 1 (Plate – 55)

This is a small form, with branches arising mostly from the basal part, 3.0 to 4.0 cm tall, bushy and dark green. At each node 3 types of branchlets, 6 in each type, are produced. The first type large, with primary ray forking into 3 to 4 dactyls and 2 to 3 secondary rays, each ending into 3 dactyls. The second type medium sized with primary ray forking into 3 to 4 dactyls and 1 secondary ray ending into 3 dactyls; the third type small with primary ray ending into 3 dactyls.Isotype at author's laboratory and Holotype No. 65 at B.S.I., CBE, T.N.

b. Form - 2

This form is more or less similar to the first one but taller and lesser bushy. This form and the previous form have no sex organs. Isotype No. 61 at B.S.I., CBE, T.N.

c. Form - 3 (Plate - 56)

This is a larger plant, more crowded and bushy, 6 to 7 cm tall, dark green. The branching pattern of larger whorl is similar to first and second forms but one or two of the tertiary rays produced from the secondary ray. Similarly the medium sized whorl has one tertiary ray from the secondary ray. The small whorl has one secondary ray produced from each primary ray. This has sex organs.Holotype No. 62 at B.S.I., CBE, T.N.

d. Form - 4 (Plate - 57)

This is the tallest plant of all the 4 forms studied but lesser crowded than the third form. The branching pattern is also more specialized, 8 to 9 cm tall; brownish green. Sex organs present.

The study of this species is a first record for KarnatakaState. Isotype No. 61 at B.S.I., CBE, T.N.

8. *Nitella tenuissima (Desv.) Kutz., cm.*

Plants small, 4 to 5 cm tall, light green, rarely branched, 5 branchlets at each node of axis, primary ray producing 3 or 4 secondary rays and a dactyl and each secondary ray ending into 2 or 3 dactyls; dactyl 2 or 3 end cells; antheridium light green and oogonium greenish brown; sex organs at the first nodes of branchlets and rare.

9. A form of *Nitella furcata* Ag.em.

A form of *Nitella furcata* Ag.em. has also been collected and studied and this will be published in a journal latter.

10. *Nitella congesta var. hyalinoides var. nov.* D. Subramanian (Plate - 58)

Isotype No. 32 kept at author's laboratory; Lectotype Sheshadri R.Holotype at B.S.I., Coimbatore, Tamil Nadu.

Plants medium sized, 5 to 6 cm tall, branched intermittently, brownish green; button like globular whorls on the axis; 6 branchlets at each axial node; each branchlet has a primary ray forking into 3 or 4 secondary rays each secondary ray into 3 tertiary ray ending into 2 or 3 dactyl; dactyl 2 cell, tip cell short and conical. All branchlets equal, shorter, crowded with bushy appearance. Only oogonium seen through out the plant.

This plant resembles *N. congesta* A.Br. in general morphology of the plants, but the branchlets are totally different in organisation. In *N. congesta*, (Icone 351, Figure 8) of Wood and Imahori, axial nodes have numerous small branchlets between larger ones and one oogonium present replacing a small branchlet. But, in the present plant, all the branchlets are of the same and uniform types and no oogonium observed at axial nodes. In *N. congesta* (Figure – 9), there is continuous percurrent central ray. But, in the present species no such percurrent central ray and the secondary rays, tertiary rays and dactyls are totally different from those of *N. congesta*. At present this plant is a distinctvariety of *N. Congesta* namely *N. Congesta* var. *hyalinoides* var nov.

In thallus organisation, this species resembles to some extent, *N. furcata*, but the primary rays in the present species is so much shortened as to make it resembling *N. congesta, N. plumosa*of Wood and Imahori and some of the forms of *N. hyalina* from Bangalore. *N. plumosa* A.Br. is dioecious, both fertile and sterile whorls similar, dactyls (1-2) 3-4 celled, the tertiary rays and dactyls curved inwards. The present plant may be also dioecious but only female plants seen and the dactyls have 2 cells and tertiary rays and dactyls are straight. The thallus of this species are all equal, unlike the unequal thallus organisation of *N. hyalina*.

Plants collected on 07.05.2015 at Gubbi of Karnataka State some 90 km away from Bangalore. This species was present alongwith *Chara bauerii* described above (Plant No. 1) in a small ditch with 1 foot depth of fresh water.Holotype No. 64 at B.S.I., CBE, T.N.

The presence of *Chara bauerii* (3 forms), *C. leptopitys, Nitella bastinii, N. hyalina* (4 forms) and *N. congesta* var *hyalinoides* is very interesting and excepting *N. hyalina*, all the other species have not been observed in Tamilnadu for the past 15 years. During these months, April and May, no Charophytes could be collected from Western Ghats, as the ditches and water reservoirs are dry there in this season.

From the Herbarium specimens deposited at DDE Wing, Botany Department, AnnamalaiUniversity, Tamil Nadu, an interesting species of *Chara* was isolated and later, it was identified, and described with illustrations by the author as *Chara caneseens* Desv & Lois. It resembles the type species described by Anders Langangen of Oslo, Norway in 2000. This species was collected from Amblical, Kolar District of Karnataka State.

11. *Chara setosa* f. *recurvata forma* nov. D. Subramanian

Plants from Herbarium specimen of DDE Wing, Botany Department, AnnamalaiUniversity, Tamil Nadu. This plant was collected in 1994.

Plants large, 12 to 14 cm tall, rarely branched, erect, branchlets of whorls reflexed and facing downwards or on sides, rudimentary stipulodes, bracts and bracteoles present, whole plant corticated. In the branchlets, thin and thick segments intermittently present, 10 segments in each branchlet; 12 branchlets in a whorl; sex organs in the lower 3 or 4, sometimes 5 nodes of segments present, in each node, one globule and one nucule present.

It resembles *C. setosa f. pseudobrachypus* (J.Gr. & Steph) R.D.W. (Icone 89) but number of branchlets in this plant is lesser, that is, 8 to 9 and some of them having 2, 3 or 4 segments instead of the usual 10 segments and this species having downwardly pointing spine cells.Holotype No. 57 at B.S.I., CBE, T.N.

Mr. Sheshadri, R. has recently sent the author(29.05.2015) some Charophytes collected from neighbouring areas of Bangalore, Karnataka State. They are described below:

12. *Lychnothamnus barbatus* f. *minima forma* nov. D. Subramanian (Plate - 59)

Plants robust, thallus incrusted, thick green, 8 to 10 cm tall. Axis stout, 800 to 900 µm in diameter, internodes c.2 x branchlet length, to 21 cm long, cortex absent, spine cells absent; stipulodes in 1 tier, 2x branchlet number, acuminate, branchlets 8 in a whorl, 2 to 2.2 cm long, segments 4, ecorticate, primary segment elongated, usually more than ½ x entire branchlet length, each node bearing 4 bract cells, verticillate, bracteoles absent, sex organs conjoined lying side by side, present at all branchlet nodes. Oogonium larger, peculiarly shaped, brownish orange, coronula 5 cells in 1 tier, convolutions 10, oospores thick, brownish orange, Antheridia small 330 to 360 µm in diameter, orange.

This plant was collected at a village Melur, some 130 km from Bangalore alongwith 2 forms of *Nitella lhotzskyi*, 3 forms of *Chara setosa* and a form of *C. fibrosa* in a small lake on 13.05.2015 by Mr. Sheshadri, R. of Bangalore.

This plant is the shortest and smallest of all the forms and varieties of *Lycnothamnus barbatus* of the world so far studied. Therefore, this plant is named herein as *L. barbatus.* f. *minima forma.* nov. D. Subramanian.

Pundhir and Aruna Gautam (1995) have given a comparative account of *Lycnothamnus barbatus. f. barbatus* (Wood and Imahori, 1965), *L. barbatus.* f. *iyengarii* (Sarma and Khan, 1966), *L. barbatus* var. *gigantii* (Bharati and Chennaveeraiah, 1983), *L. barbatus. f.giganteum* (Patel and Jawala, 1985) and *L. barbatus f.mahmoodia* (Pundhir and Aruna Gautam, 1995), including qualitative and quantitative differences of morphological characters among them. When compared to those forms and varieties of *L. barbatus* the present plant is different in larger number of characters and so it is named herein as a new form.Isotype deposited at author's home laboratory and Holotype No. 67 at B.S.I., CBE, T.N.

13. *Nitella verticillata* (Fil & G.O.A, ex. Fil.) R.D.W. (*Charina verticillata* Fil & G.O.A. ex. Fil.) (Plate - 59)

Wood and Imahori (1965) have illustrated the plants of condensed form and lax form of this species and their parts but no description of them were provided. This plant has no sex organs (Icone 376). The author has already included a form of this species from "FairyFalls" of Kodaikanal with sex organs and another form from Sathamangalam of Virudhachalam Taluk in his book from Tamilnadu.

Plant 8 to 9 cm tall, branched, erect, internode 2 times the length of branchlet whorls, 5 branchlet at each node of axis, primary ray forks into 3 to 4 dactyls and a percurrent secondary ray ending into 3 to 4 dactyls; dactyls 2 celled; Gametangia absent.

This is a small form of this species from KarnatakaState. Herbarium specimens are deposited at author's house laboratory at 277, Mariappanagar, Annamalai Nagar – 608 002, Tamil Nadu. Holotype No. 68 at B.S.I., CBE, T.N.

14. *Nitella lhotzkey* (A.Br.) A.Br., em. Form I (Plate - 60)

Plants olive green, 5 to 7 cm tall, branched freely, internodes 2 times longer than branchlet whorls. Two types of branchlets, 8 larger and fertile and 8 smaller and sterile, primary ray longer, about ¾ of the entire length of branchlets, ending into 2 to 3 dactyls and 2 to 3 secondary rays, each finally ending into 2 to 3 dactyls; dactyl 2 celled; tip cell conical, each sterile branchlet ends into 2 to 3 dactyls. Sex organs at the first nodes of fertile branchlets. One oogonium and one antherium at each node.

This plant differs from other forms of this species so far studied in having comparatively longer primary rays of fertile branchlet whorls. Herbarium specimens are deposited at author's house laboratory.

15. *N. lhotzkeyi* (A.Br.) A.Br. em. Form - 2 (Plate - 61)

This plant is larger than the previous plant of this species in all dimensions but primary rays of larger whorls are ½ of the entire length of branchlet whorls. Plants 8 to 10 cm tall and dactyls are broader and longer than those of previous plant. Sex organs are present at the first nodes of branchlets. Antherium is light green and oogonium light orange. Holotype No. 63 at B.S.I., CBE, T.N. Isotype at author's lab.

16. *Chara fibrosa* (one form) and *Chara setosa* (three forms)

Chara fibrosa (one form) and *Chara setosa* (three forms) have also been collected in this place along with already described plants.

The Charophytes of Karnataka state are more or less similar to those of North India but mostly different from those of Tamil Nadu. The species like *Chara canescens, C. bauerii* and *C. leptopitys* and *Nitella bastinii, N. duthiae* and *N. sheshadrii* have not been so far collected from Tamilnadu, eventhough they are commonly available in Karnataka state. From Kerala, it is not possible to comment on the floristic nature of Charophytes, as no intensive study on this group of plants has been made so far. Similar is the case with Andhra Pradesh.

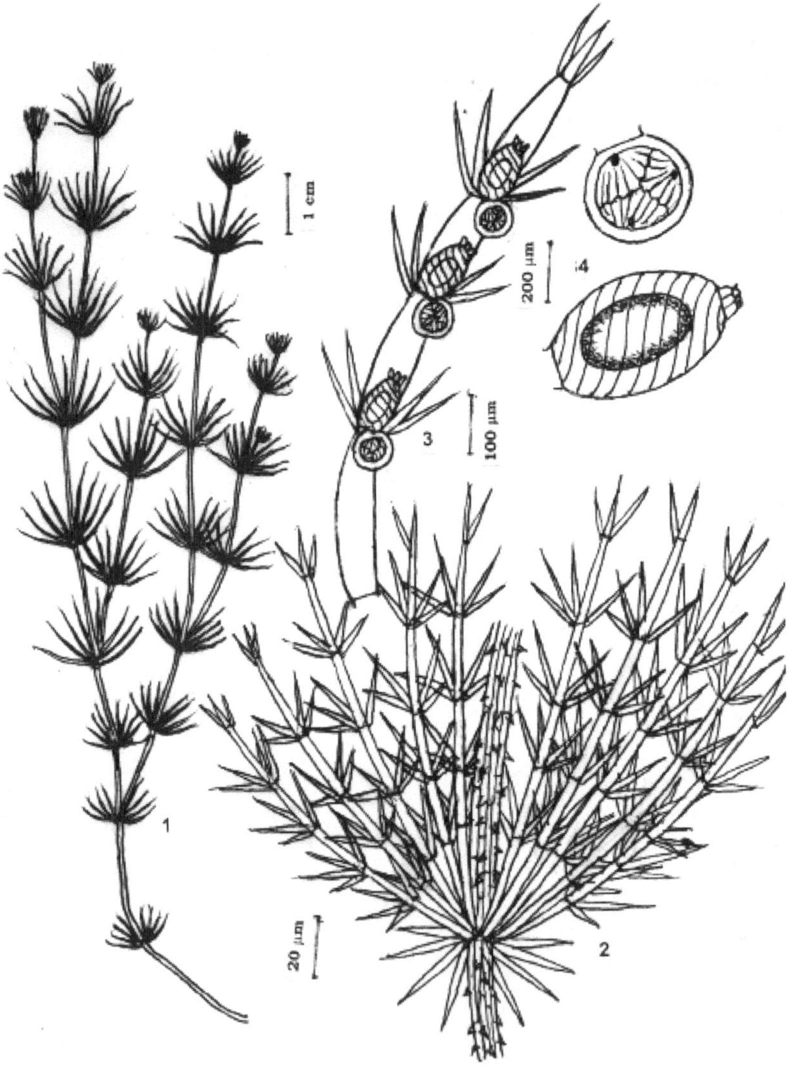

Plate 49 *Chara bauerii*. var. 1. Fig. 1 = A plant; 2 axial node with branchlets; 3 – 4 sex organs.

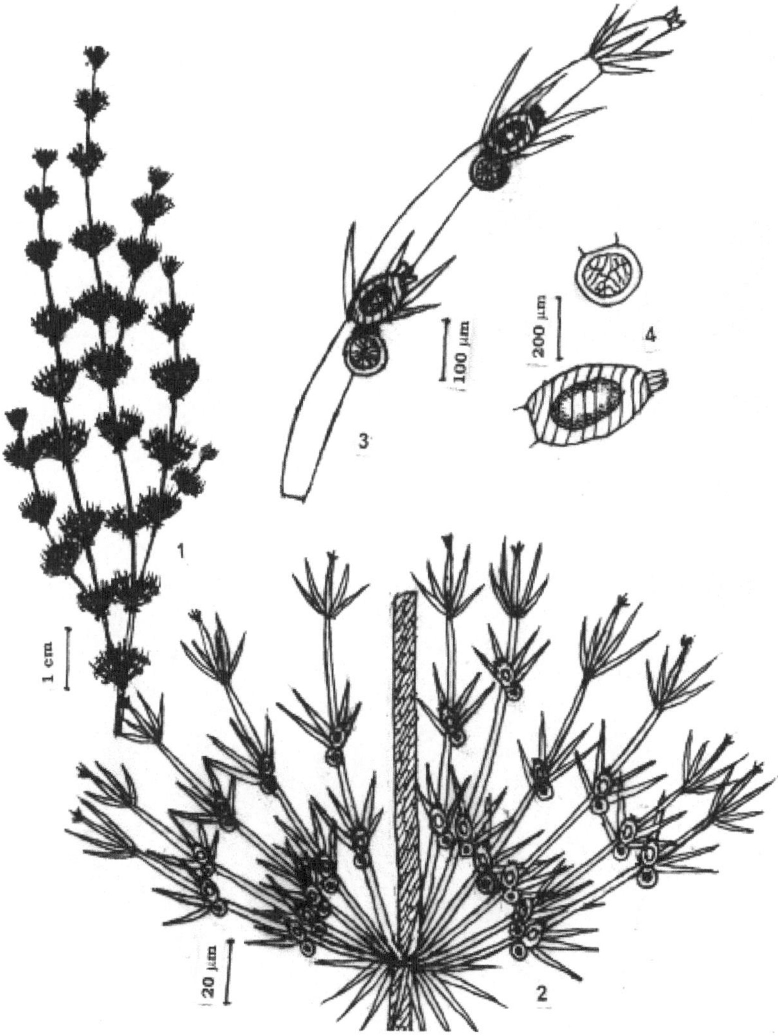

Plate 50 *Chara bauerii*. var. 2. Fig. 1 = A plant; 2 axial node with branchlets; 3 = A branchlet enlarged; 4 = Sex organs.

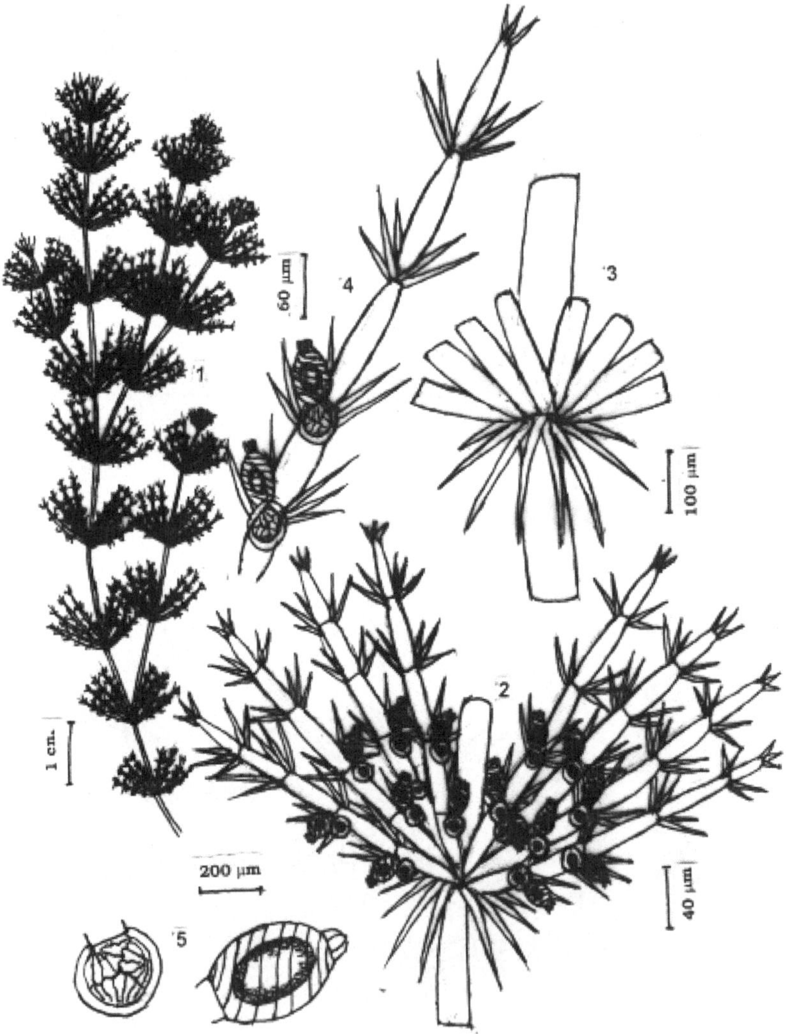

Plate 51 *Chara bauerii*. Var 3 Fig. 1 = A plant; 2 = Axial node with branchlets; 3 = Axial node with stipulodes; 4 = A branchlet; 5 = sex organs.

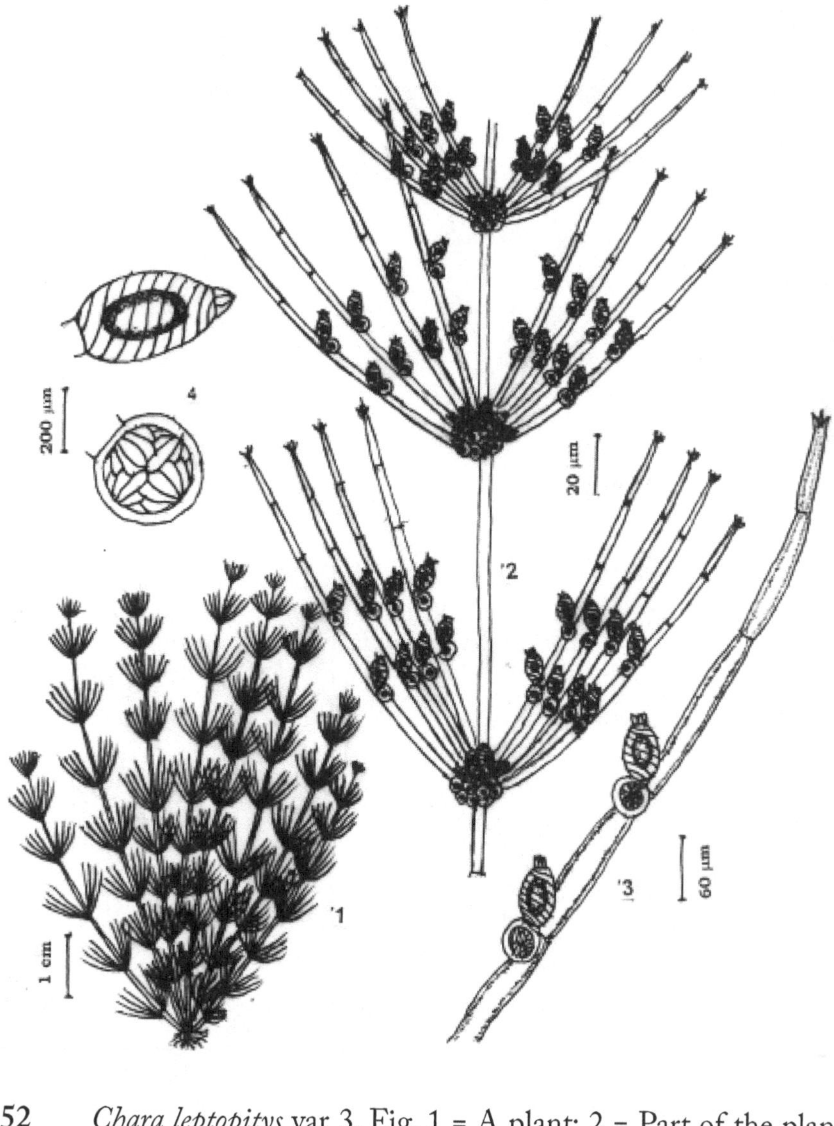

Plate 52 *Chara leptopitys* var 3. Fig. 1 = A plant; 2 = Part of the plant enlarged; 3 = A branchlet with sex organs; 4 = sex organs.

Plate 53 *Nitella bastinii* Fig. 1 = A plant enlarged; 2 = A fertile whorl; 3 = A fertile branchlet with sex organs; 4 = Dactyls of fertile branchlet; 5 = dactyls of sterile branchlet.

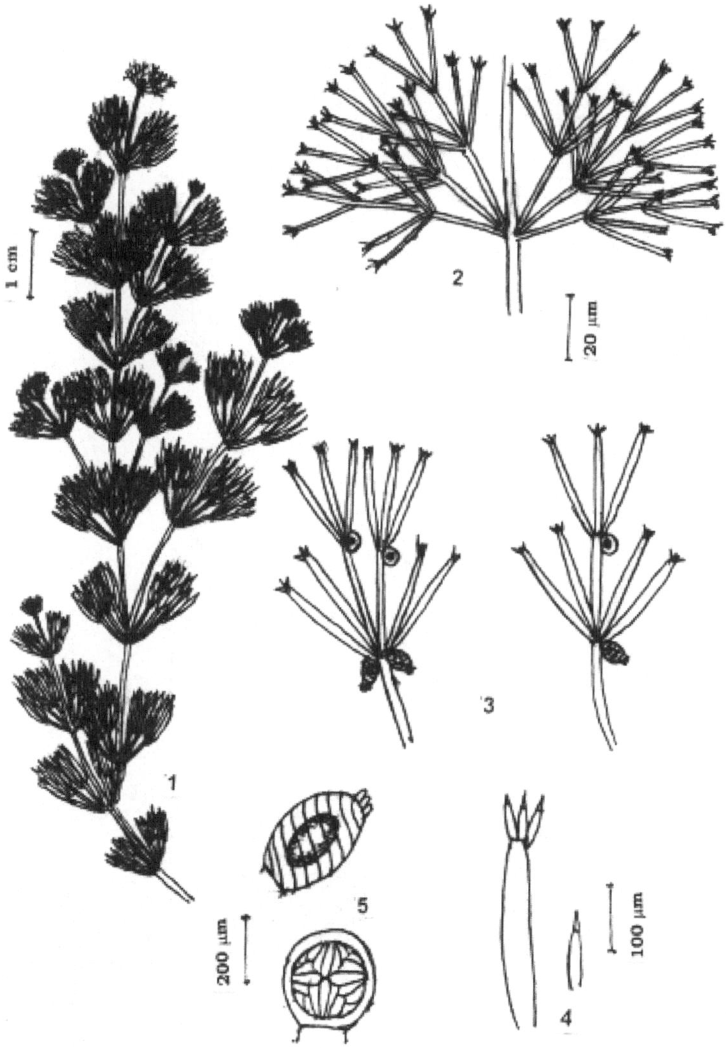

Plate 54 *Nitella duthieae*. Fig. 1 = A plant; 2 = Tip whorls; 3 = Branchlets; 4 Dactyls; 5 = sex organs.

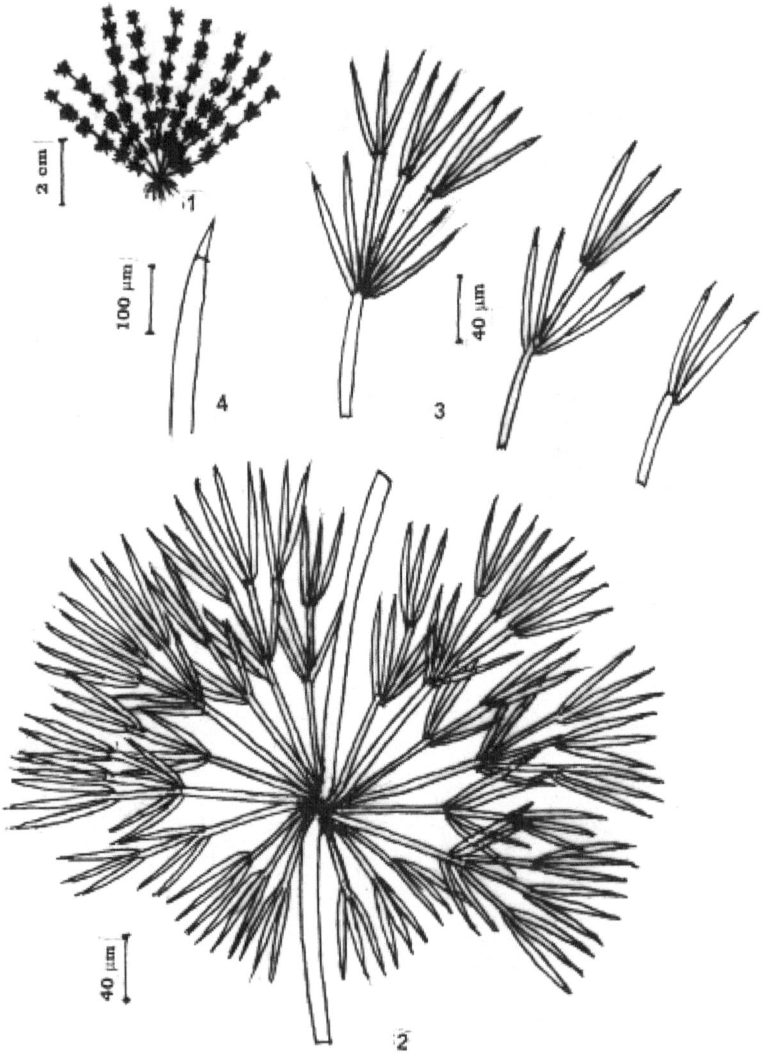

Plate 55 *Nitella hyalina*, Form 1 Fig. 1 = A plant; 2 = A whorl enlarged; 3 = 3types of branchlets; 4 = A dactyl

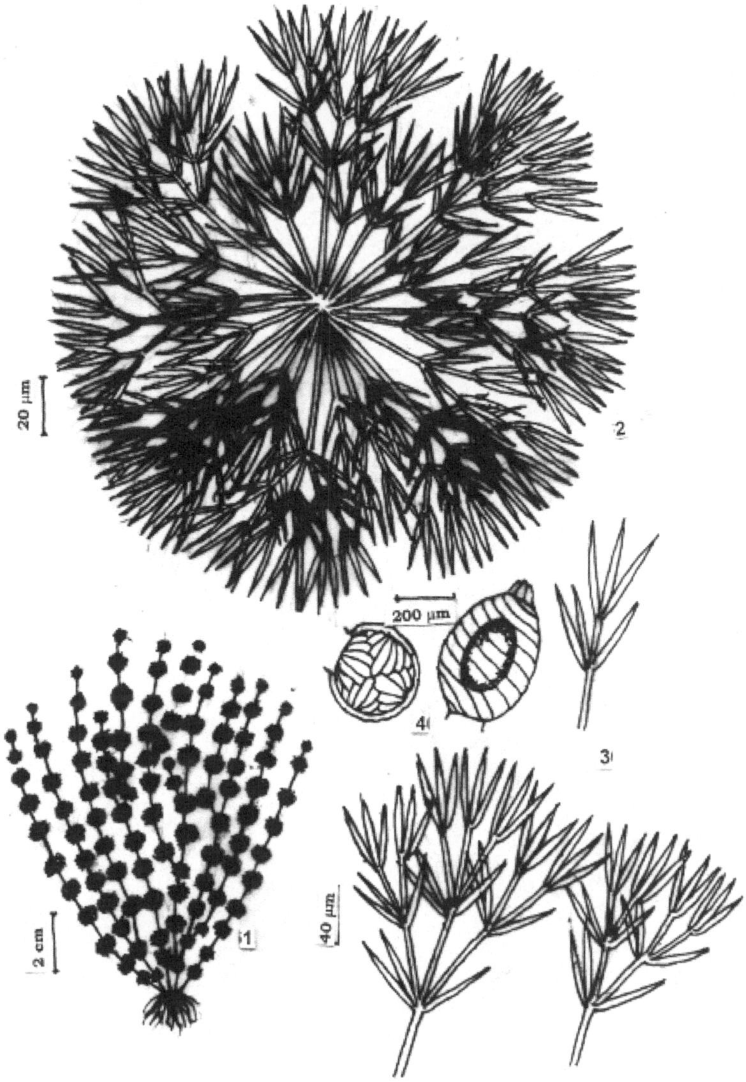

Plate 56 *Nitella hyalina*. Form – 3. Fig. 1 = A Plant; 2 = A whorl enlarged; 3 = 3 types of whorls; 4 = sex organs.

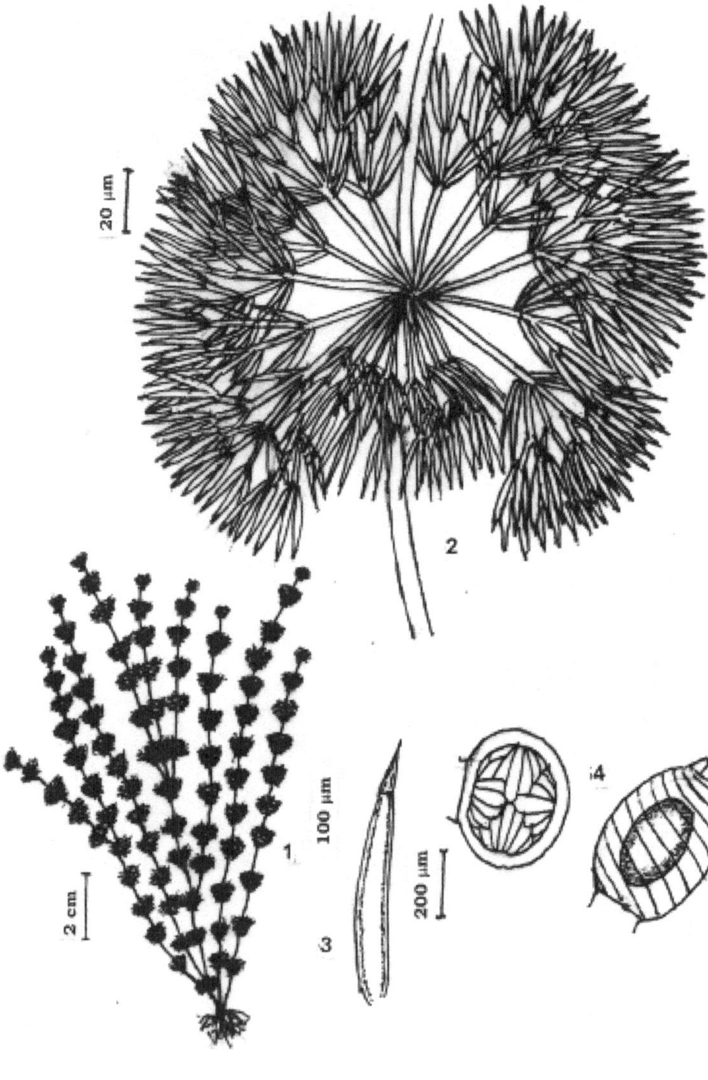

Plate 57 *Nitella hyalina* Form – 4 Fig. 1 = A plant; 2 = A whorl enlarged;

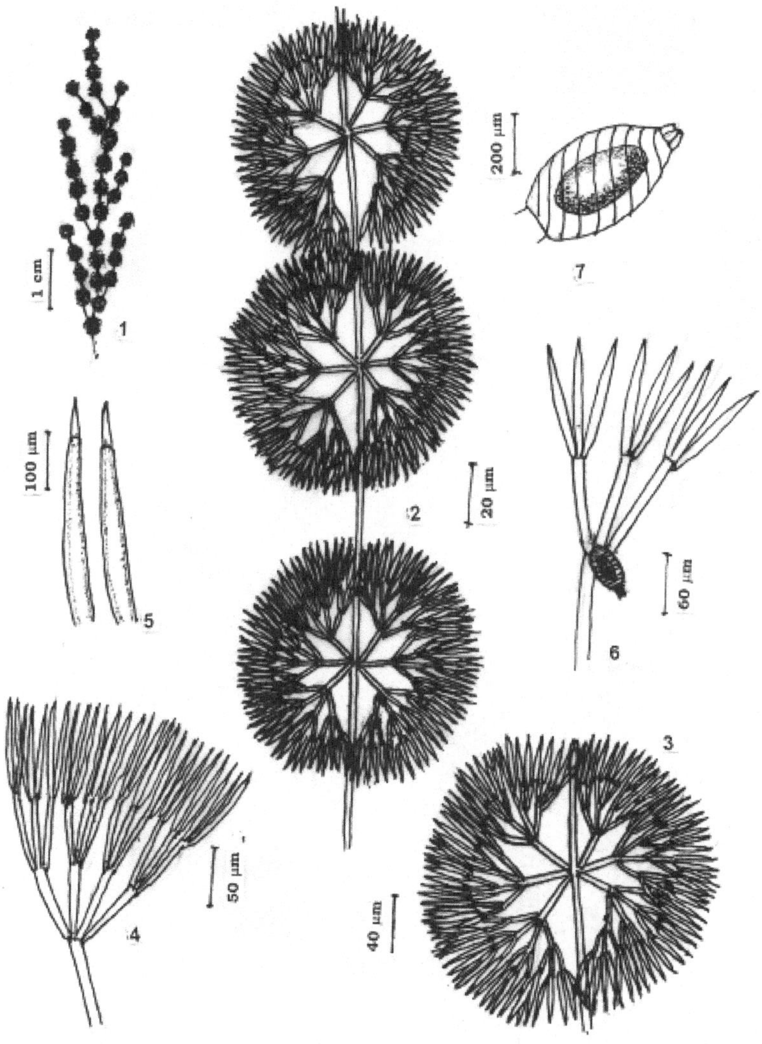

Plate 58 *Nitella congesta* var. *hyalinoides* Fig. 1 = A plant; 2 = Portion of plant enlarged; 3 = A whorl; 4 = A branchlet ; 5 = A dactyl; 6 = part of whorl with 100 gonium 7 = Oogonium.

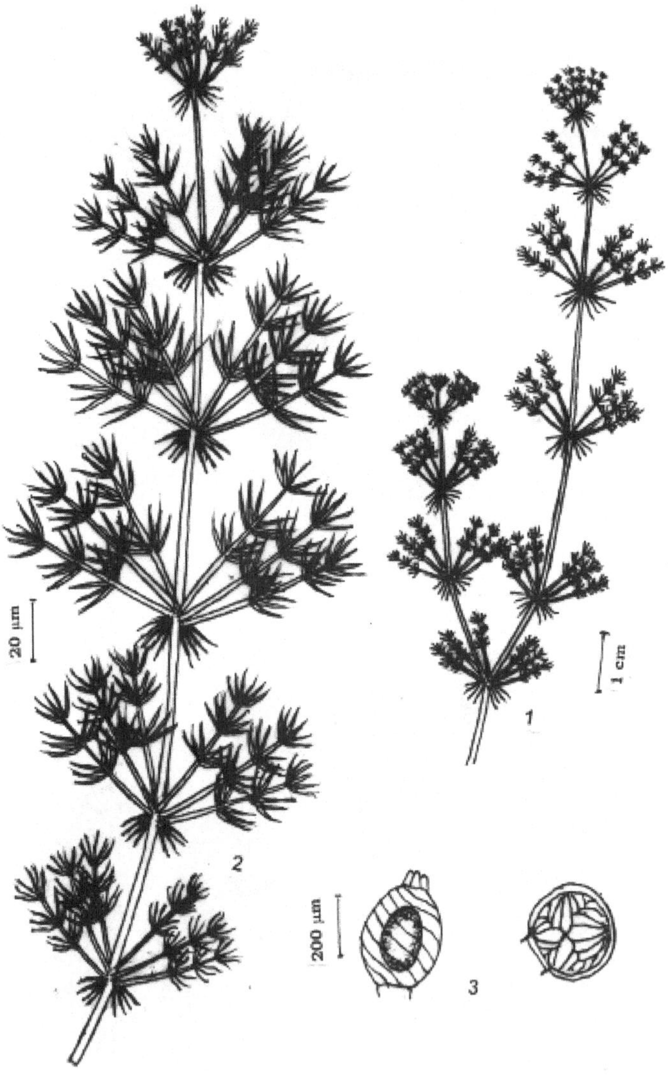

Plate 59 *Lychnothamnus barbatus. f. minima.* Fig. 1 = A plant slightly enlarged; 2 = A plant more enlarged; 3 = sex organs.

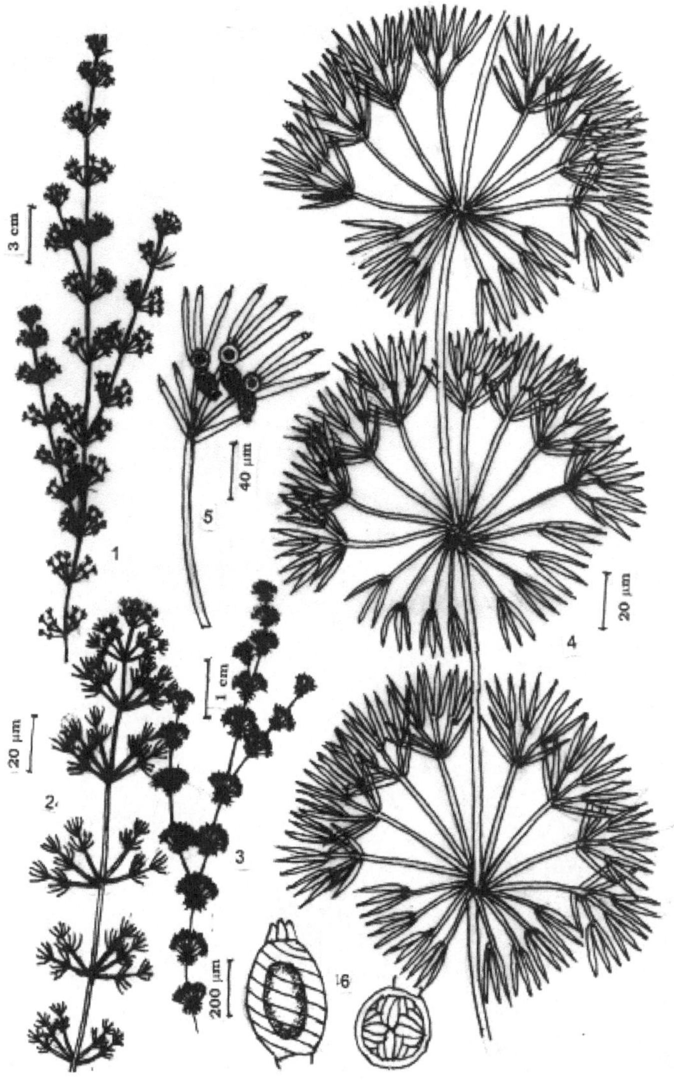

Plate 60 *Nitella lhotskeyi* form – 1

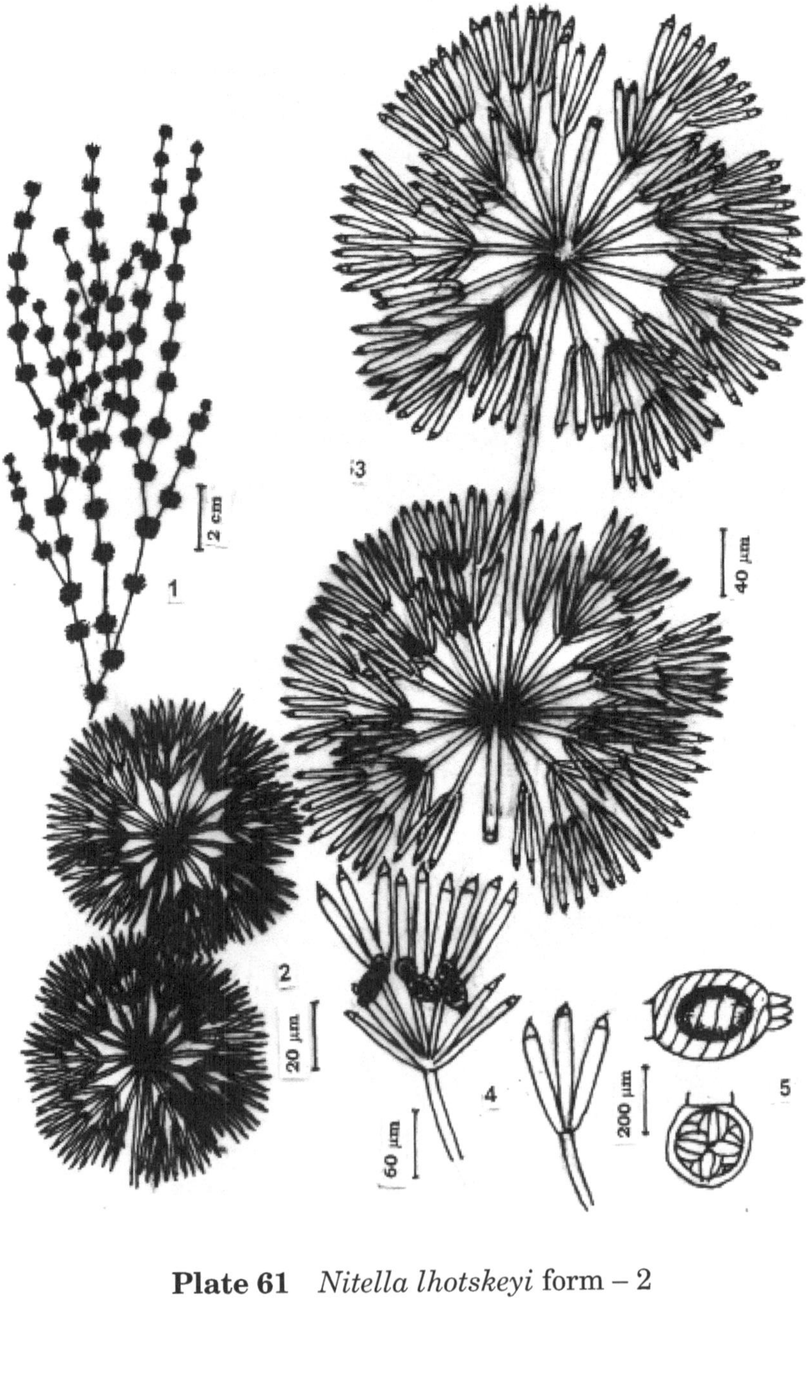

Plate 61 *Nitella lhotskeyi* form – 2

10. CHAROPHYTES STUDIED FROM NORTHERN PARTS OF INDIA

1. *Chara globularis* Thuill, em. (Plate - 62)

This plant was collected from Patiala, Punjab. Herbarium specimens are deposited at DDE Wing, Botany Department, AnnamalaiUniversity, Tamil Nadu.

Plants medium sized, 10 to 12 cm tall, rarely branched, main stem and branchlets corticated excepting the 2 uppermost segments being ecorticated; 8 branchlets at each node, branchlet 8 segmented, no bracts, bracteoles, stipulodes and sex organs noted.

In *Chara globularis f.capensis* R.D.W. at each node of branchlets, 5 to 6 rudimentary bracts cells are present. But in the present species the bracts cells are dotted circular, at all the nodes of branchlets. In the former form, there are 8 segments in each branchlet but in the present plant, there are 7 segments. Morphological appearances of the two plants are almost similar.

2. *Chara corallina* Klein ex. Willd em. (Plate 63)

This plant was collected from Ambala of North India. Vaucher specimens have been deposited at DDE Wing, Botany Department, AnnamalaiUniversity, Tamil Nadu.

Plants taller, 20 to 24 cm high, rarely branched, almost erect, stem and branchlets all ecorticated, 6 branchlets at each node of main axis, each branchlet 6 segmented, at each node of branchlets circular cells present, no stipulodes, bracts and bracteoles, sex organs not seen.

This plant closely resembles *Chara corallina.f. kyusyensis* R.D.W. in almost all the characters but the plant is sterile.

3. *Chara vulgaris f. himalayensis forma* nov. D. Subramanian

This plant was collected in 1991 from Surinsar of Jammu. Herbarium specimens deposited at DDE Wing, Botany Department, Annamalai University, Annamalai Nagar – 608 002, Tamil Nadu.

Plants whitish green when dry, tall, 20 to 22 cm, rarely branched, erect, branchlet whorls longer than internodes. 8 branchlets in an axial node, 9 to 10 segments in each branchlet, no spine cells, all branchlets and main axis corticated, no stipulodes, bracts or bracteoles. Segments of branchlets thicker and thinner alternating in irregular pattern; sex organs only at the basal nodes of branchlets, one antheridium and one oogonium at each node.

This plant resembles *C. vulgaris f.pistianensis* (Vilh.) R.D.W. (Icone 6) in the absence of bracts, bracteoles and presence of cortication of axis and branchlets but differs from it in having more number of segments of branchlets being 9 to 10, instead of 7 to 8 in the above form of this species. In the present form no stipulodes observed but in *C. vulgaris f.pistianensis* rudimentary stipulodes are present in 2 tiers. Sex organs are present in the basal first nodes of branchlets in the present plant but they are present in the basal 3 nodes of each branchlet in *C. vulgaris. f. pistianensis*. Therefore, the present plant is named as a separate form. Isotype No. 69 at B.S.I., CBE, T.N.

After the publication of the author's book "Monograph on Indian Charophyta" in 2002, a large number of species have been morphologically and cytologically studied from various parts of India; particularly from North India.

They are given already in the introductory part of this book.

Charophytes are peculiar, delicate and complicated plants. Mere descriptions of a plant in whatever details may be, will not help to identify and understand the status of it. That is why, in Wood and Imahori's book, most of the charophytes described were drawn and provided as an equally important "Iconograph" as descriptions of them. The author is taking steps to illustrate as many members of Indian Charophyta as possible, so that those who will be interested in the study of them in the future may not find any difficulties in this attempt.

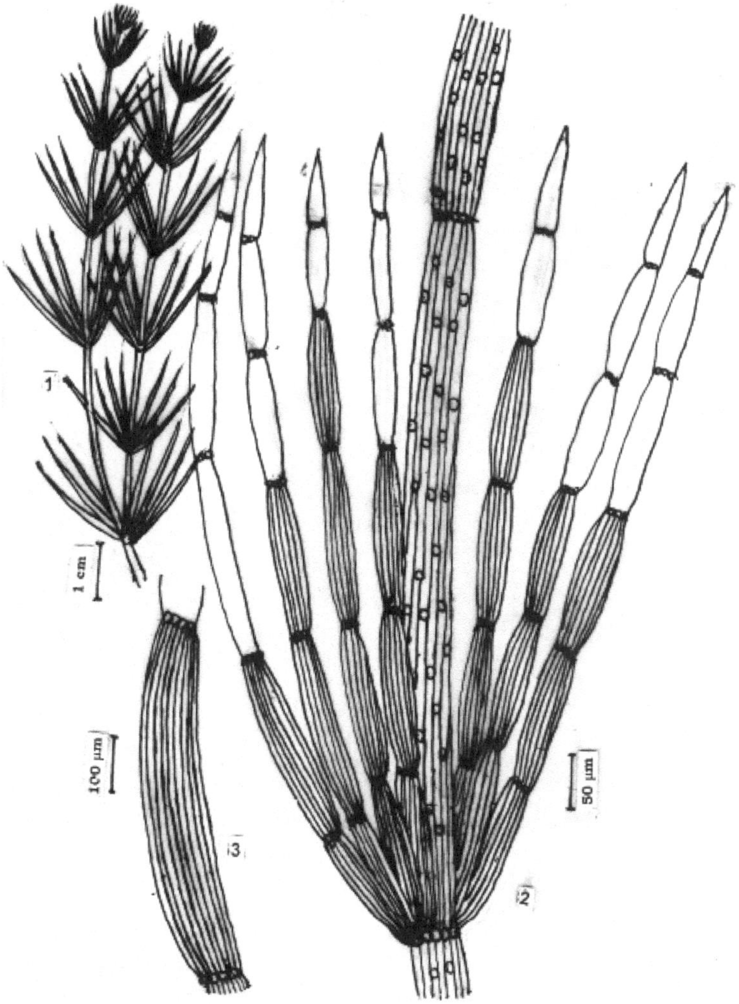

Plate 62 *Chara globularis. From patiala* Fig. 1 = A plant; 2 = A portion of the plant with main stem and a whorl of branchlets enlarged. 3 = A branch with 2 nodes enlarged.

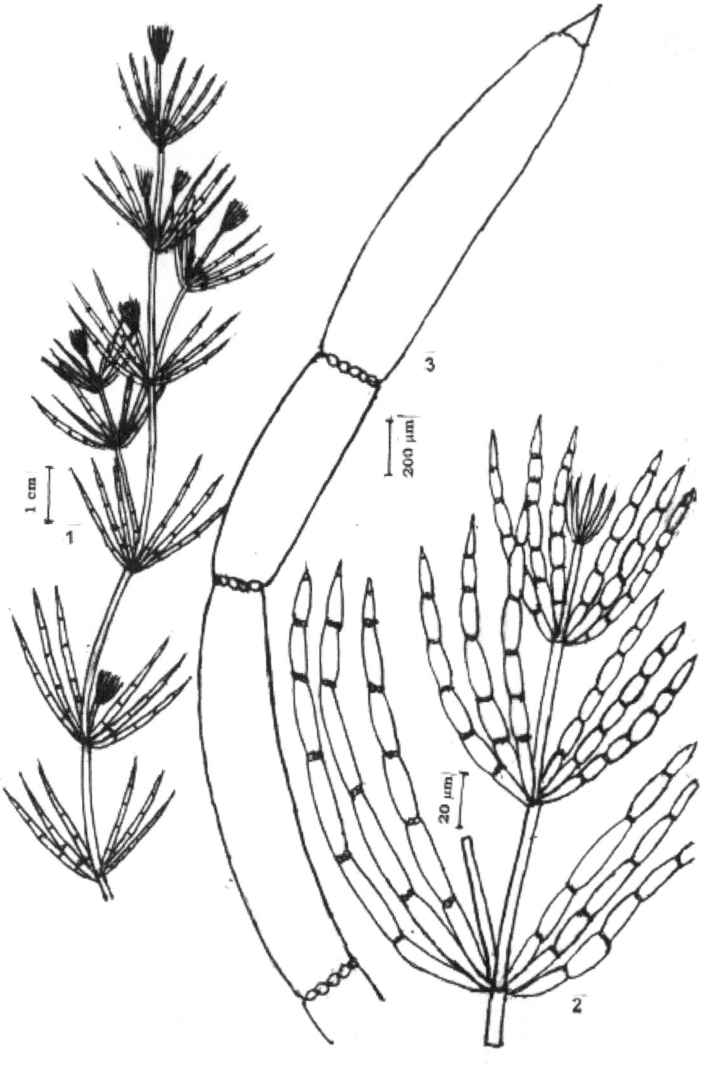

Plate 63 *Chara corallima* from Ambala

REFERENCES

1. Alston, R.E. and Turner, B.L., 1963. "*Biochemical Systematics*", Prentice Hall, Inc., New York.

2. Bhatnagar, S.K., 1981. *Cytotaxonomy of Charophytes of Rohilkhand division and effect of different chemicals on selected taxa.* Ph.D. Thesis, RohilkhandUniversity, Bareilly.

3. Bhatnasagar, S.K., 1983. The concept of basic chromosome numbers in Charophytes – A review. *Cryptogamie, Algologie* 4, 111-116.

4. Bhatnasagar, S.K., 1987. Phylogeny and interrelationships of the genus *Tolypella*: A multidisciplinary outlook. *Phykos* 26, 140-148.

5. Bhatnasagar, S.K., 1988a. *Nitella hyalina f. hyalina* (DC) Ag. RDW : A new chromosome count and its cytotaxonomy. *Current Sci.* 57: 343-344.

6. Bhatnasagar, S.K., 1988b. The first cytotaxonomic report in *Chara braunii f. novi-mexicana* (Div. Charophyta). *Current Sci.* 57, 88-90.

7. Bhatnasagar, S.K., 1988c. A karyological approach to phylogeny and interrelationship of Charophyta, *Cytologia* 53, 723-729.

8. Bhatnasagar, S.K., 1989a. Cytological perspectives of Charophyta. I. *Chara globularis* complex, *Phykos* 28, 156-165.

9. Bhatnasagar, S.K., 1989b. Cytological perspectives of Charophyta. II. *Nitella furcata* complex, *Phykos* 28, 166-177.

10. Bhatnasagar, S.K., and L.P. Gangwar, 1985. On *Tolypella intricata* (Trent ex Roth) Leonh, em RDW, Characeae from Indian sub-

continent, A karyomorphological approach, *Cryptogamie. Algologie* 6, 185-190.

11. Bhatnasagar, S.K., and M. Johri, 1986. Some aneuploids in Indian Charophyta, *Acta. Bot. Neerl* 35, 377-381.

12. Bhatnasagar, S.K., and M. Johri, 1991. Preliminary investigations on cytology and cytotaxonomy of Charophytes from Rohilkhand division, Bareilly, *J. Indian Bot. Soc.,* 70, 55-58.

13. Bhatnasagar, S.K., and M. Sharma, 1985. A new chromosome numbers in the genus *Nitella* from India, *Sci. and Cult.* 51, 195-196.

14. 13a. Barathi, S.G and Chennaveeraiah, M.S. 1980. Cytotaxomical Studies on *Chara zeylanica* complex in "Taxonomy of Algae" *International Symposium,* University of Madras, 353-365.

15. Buzzati 'Traverso, A.A., 1969. *Paper chromatography in relation to Genetics and Taxonomy. New approaches in cell Biology,* Academic Press, New York.

16. Chatterjee, P., 1970. *Lychnothamnus barbatus* (Meyen) Loenh in West Bengal and its Chromosome number, *Bull bot Soc. Bengal,* 24, 131-133.

17. Chatterjee, P., 1971. *An analysis of the structure and behaviour of chromosomes as an aid in the taxonomy of certain higher and lower groups of plants and the cytological effect of certain chemical and physical agents in the latter.* Ph.D. Thesis, CalcuttaUniversity, Calcutta.

18. Chatterjee. P, 1975. Some additions to the Charophytes of West Bengal, *Bull. Bot. Soc.* Bengal 29, 105-109.

19. Chatterjee. P, 1976a. Cytotaxonomic studies on *Chara socotrensisNordst. f. nuda* (Pal) RDW from West Bengal, India. *Cytologia* 41, 659-663.

20. Chatterjee. P, 1976b. Cytotaxonomical studies of West Bengal Charophytes: Karyotype analysis in *Chara braunii, Hydrobiologia,* 49, 171-174.

21. Chatterjee. P, 1979a. *Chara fibrosa var. fibrosa f. longicorollata* : A new record for India and its cytology, *Current Sci.* 48, 545-547.

22. Chatterjee. P, 1979b. Karyological investigation of *Chara wallichii* A Br from West Bengal. *Cell Chromosome Newsl* 2, 21-23.

23. Chennaveeraiah M.S. and S.G. Bharati, 1974. Morphological and cytological observations in *Chara gymnopitys* A. Br.*Cytologia*, 39, 443-451.

24. Guerlesquin Micheline, 1960. Contribution a 'b' c tude chromosomique des charophycees d' Europe occidentale et d arbique du nord. *Rev. Gen. Bot. 68* (805) : 360-372.

25. Harborne, J.E., 1973. *Phytochemical Methods* (A Guide to modern techniques of plant analyses). Chapman and Hall, London, Toppan Company Ltd., Tokyo, Japan.

26. Hotchkiss, A.T., 1956. Some chromosome numbers in Kentucky Characeae. *Trans Kentucky Acad. Sci.* 19(1-2) : 14-18.

27. Khan, M., 1966. *Studies on cytology and cytotaxonomy of Indian Charophyta,* Ph.D. Thesis, BanarasHinduUniversity, Varanasi, India.

28. Khan, M.,1991. Charophytes in time and space: Zonal distribution pattern. *Bull Soc. Bot.* Fr 138, *Actual Bot.* 1, 33-34.

29. Khan, M., and Y.S.R.K. Sarma, 1967a. Studies on cytotaxonomy of Indian Charophyta I, *Chara Phykos* 6, 36-47.

30. Khan, M., and Y.S.R.K. Sarma, 1967b. Studies on cytotaxonomy of Indian Charophyta. II, *Nitella. Phykos* 6, 48-61.

31. Khan, M., and Y.S.R.K. Sarma,1967c. Some observations on the cytology of Indian Charophyta.*Phykos* 6, 62-74.

32. Khan, M., and Y.S.R.K. Sarma,1980. New chromosome numbers for Indian Charophyta. *Cell and Chromosome Newsletter,* 3, 24.

33. Khan, M., and Y.S.R.K. Sarma, 1981-82. Cytogeographic study of Charophyta with particular reference to India. In: *Recent Advances in Cryptogamic Botany* (ed. D.C. Bharadwaja). The Palaeobotanical Society, Lucknow (1979), 55-82.

34. Khan, M., and Y.S.R.K. Sarma, 1984. Cytogeography and cytosystematics of Charophyta In : *Systematics of the Green Algae* (eds D E G Irvine and D M John), Academic Press, London, 303-330.

35. Krishnamurthy, I.V. and Subramanian, D., 1978. Biochemical studies in *Leerzia* and *OryzaJ. Indian. Bot. Soc.* 57 : 34-39.

36. Labh, L. and B.N. Verma, 1984. Cytological studies in three species of *Tolypella, Cytologia,* 49, 717-723.

37. Labh, L. and B.N. Verma, 1985a. First report of chromosome numbers in four taxa belonging to the genus *Chara* Linn. *Cytologia* 50, 21-24.

38. Labh, L. and B.N. Verma, 1985b. New record of chromosome numbers for the Charophytes, *Cytologia* 50, 55-58.

39. Labh, L. and B.N. Verma,1986. New counts for *Chara fibrosa* complex. Cytologia 51, 185-191.

40. Mukherjee, S., 1978. Polyploid chromosome numbers in the genus *Nitella, Current Sci.* 47, 386.

41. Mukherjee, S., and M.N. Noor, 1973. A first report of Karyological studies in *Nitella wattii* Groves from India, *Sci. and Cul.* 39, 459-460.

42. Mukhopadhyay, A. and P. Chatterjee, 1980. Cytomorphology of *Chara flaccidaA. Br.* from West Bengal, *Cell and Chromosome Newsl,* 3, 37-39.

43. Noor, M.N. 1967. Chromosome numbers in two Indian species of *Chara,* viz., *C. corallina* and *C. hydropitys* Reich, *Proc. Indian Sci. Cong.* (Abstr.) 392.

44. Noor, M.N., 1968. A new record of chromosome number in *Nitella mucronata* (Br.) Miquel from India. *Sci. and Cul.* 34, 214-215.

45. Noor, M.N., 1970. A preliminary report on the chromosome numbers of some Indian Characeae, *J. Ranchi Univ.* 6, 7, 233.

46. Noor, M.N. and S. Mukherjee, 1974. Karyology of *Nitella superba* from India. *Proc. Inter Symp. Taxo.* Algae Madras, 28.

47. Noor, M.N., and S. Mukherjee., 1975. On the aneuploid chromosome number in *Chara hydropitys* Reich from India, *Cytologia* 40, 803-807.

48. Noor, M.N., and S. Mukherjee., 1977. Some new records of chromosome number in Indian Charophyta, *Cytologia* 42, 227-232.

49. Pal, B.P., B.C. Kundu, V.S. Sundaralingam and G.S. Venkataraman, 1962. *Charophyta,* I.C.A. R. Publn., New Delhi, India.

50. Pal, R. and P. Chatterjee, 1986a. Feulgen cytophotometric determination of nuclear D N A in species of *Nitella* (Charophyta) from India. *Cryptogamie Algologie* 129-134.

51. Pal, R. and P. Chatterjee, 1986b. Feulgen cytophotometric determination of nuclear D N A in species of *Chara* (Charophyta) from India, *Cryptogamie, Algologie,* 7, 317-322.

52. Pundhir, H.S. and Aruna Gautam, 1993. Occurrence of *Nitella hyalina* var. *hyalina f. brachyactis* (A. Br.) Fldm. R.D.W. a new record for India. *Phykos* 32 (1&2) : 99-104.

53. Pundhir, H.S., Aruna Gautam and Vidyavati, 1993. Comparative study and new chromosome counts for *N. hyalina* var. *hyalina f. indica* (A.Br.) R.D.W. *Phykos.* 32 (1&2): 169-174.

54. Pundhir, H.S. and Vidyavati, 1993. Taxonomic enumeration and chromosome counts of some charophytes from district Aligarh, Utter Pradesh, *J. Mendel,* Vol. 10 (2-4) : 83-85.

55. Pundhir, H.S. and Vidyavati, 1994. Some new Karyomorphological observations of *Chara fibrosa.* var. *fibrosa. f. longicorollata* (Kasaki) R.D.W. *J. Indian Bot. Soc.* 73 : 245-249.

56. Pundhir, H.S., Aruna Gautam and Vidyavati, 1994. Some charophytes from Aligarh district, Uttar Pradesh, *Phykos.* 33(1&2): 1-9.

57. Pundhir, H.S. and Vidyavati, 1994. New karyomorphological observations on Charophytes from Uttar Pradesh, India. *The nucleus* 37(1&2) : 39-45.

58. Pundhir, H.S. and Aruna Gautam, 1994. Observations on morphology and Cytology of *Chara vulgaris* var. *vulgaris f. crispa* (Wallon) R.D.W. A new record for India*Current Researches in Plant Sciences,* pp. 197-201.

59. Pundhir, H.S. 1994. Studies on the karyomorphology of *Chara hornemannii f. longifolia* (Rob.) R.D.W. A new record for India. *J. Mendel*-II (3-4) : 157-160.

60. Pundhir, H.S. and Aruna Gautam, 1995. Cytomorphological studies on a new form of *Lychnothamnus barbatus f. mahmoodia f.* nov. *Phykas* 34(1&2) : 39-48.

61. Pundhir, H.S., Neeta Chauhan and A.K.S. Chauhan, 1999. Studies on cytology and cytotaxonomy of *Nitella duthieae* (Charophyta : Charophyceae) : A first record from India. *J. Cytol. Genet.* 34(1) : 45-48.

62. Pundhir, H.S. and A.K.S. Chauhan, 1999. Chromosome number in the alga *Chara socotrensis f. socotrensis* R.D.W.; A new record for India. *Bionotes* 1(4) : 77.

63. Ramjee, 1969. *Contributions to the systematics, cytology and cytotaxonomy of Indian Charophyta.* Ph.D. Thesis, BanarasHinduUniversity, Varanasi, India.

64. Ramjee and S.K. Bhatnagar, 1978a. Studies on Charophytes from Rohilkhand division. I. Moradabad : Taxonomic enumeration and chromosome counts. *Phykos* 17 : 87-93.

65. Ramjee and S.K. Bhatnagar, 1978b. Significance of a new chromosome number in *Nitella mirabilis* Nordst ex. Groves em R.D.W. Hydrobio. 53 : 99-101.

66. Ramjee and S.K. Bhatnagar, 1978c. Cytotaxonomic studies on *Nitella furcata* subsp *flagellifera f. patula* Gr. ex Allen and comparison with other taxa of *N. furcata* complex. *Caryologia* 31 : 457-462.

67. Ramjee and Y.S.R.K. Sarma, 1971. Some observations on the morphology and cytology of Indian Charophyta, *Hydrobiol.* 37 : 367-382.

68. Ray, S. and P. Chatterjee, 1983. A new technique in karyological investigations of Characeae. *Cell and Chr. Res.* 6 : 27-29.

69. Ray,S. and P. Chatterjee, 1984. A note on the nucleolar staining in *Chara* sp.*Current Sci.* 53 : 1301-1302.

70. Ray, S. and P. Chatterjee, 1986a. *Chara vulgaris* var. *inconnexa f. hippelliana,* A new record for India and its karyotype, *Current Sci.* 55 : 662-664.

71. Ray, S. and P. Chatterjee, 1986b. Karyotype analysis in *Chara braunii* G. M.*f. braunii* (Characeae), *the Nucleus,* 29 : 33-36.

72. Ray, S. and P. Chatterjee,1987. Karyomorphological investigations of three populations of *Chara corallina* var and *f. corallina* in West Bengal, *Cytologia,* 52 : 455-458.

73. Ray, S. and P. Chatterjee,1988a. Cytotaxonomical study of some members of *Chara fibrosa* complex. *Cryptgamie Algol.* 9 : 65-71.

74. Ray, S. and P. Chatterjee, 1988b. Cytotaxonomical study of West Bengal Charophyta. A new chromosome number count of *Nitella stuartii* and its karyotype,*Acta Bot. Neerl.* 37 : 523-525.

75. Ray, S. and P. Chatterjee, 1989. A new chromosome number report for *Chara globularis* var *leptosperma f. leptosperma,Current Sci.* 58 : 312-313.

76. Sarma, T.A. and V. Bala, 1984. New chromosome counts in four taxa of *Chara* from Punjab,*J. Cytol. Genet.* 19 : 27-32.

77. Sarma, Y.S.R.K. 1958. *Cytological and cultural studies of some members of Ulotrichales and other Chlorophyceae.* Ph.D. Thesis, LondonUniversity, London.

78. Sarma, Y.S.R.K, 1968. Cytology and cytotaxonomy of Indian Charophyta – A resume. *The Nucleus* (Sem Chromosomes, Vol.) 266-285.

79. Sarma, Y.S.R.K, 1973. Algal karyology in India. In Advancing Frontiers in Cytogenetics. (*Prof. P.N. Mehra Comm. Vol., ed. P. kachroo*) : 266-285.

80. Sarma, Y.S.R.K, 1964. Chromosome numbers in some Indian species of *Nitella. Chromosoma* 15 : 246-247.

81. Sarma, Y.S.R.K and M. Khan, 1965a. Chromosome numbers in some Indian species of *Chara, Phycologia,* 4 : 173-176.

82. Sarma, Y.S.R.K and M. Khan, 1965b. A preliminary report on the survey of chromosome numbers of Indian Charophyta. *The Nucleus,* 8 : 33-38.

83. Sarma, Y.S.R.K and M. Khan, 1965c. Some new observations on the karyology of *Chara zeylanica* Klein ex Willd, *Current Sci.* 34 : 293-294.

84. Sarma, Y.S.R.K and M. Khan, 1966. A new form of *Lychnothamnus barbatus* from India. *Phykos,* 5 : 195-197.

85. Sarma, Y.S.R.K and M. Khan,1967. On two new Charophytes from India. *Hydrobiol.* 30 : 405-410.

86. Sarma, Y.S.R.K, M. Khan,and Ramjee, 1970. A cytological approach to phylogeny, interrelationship and evolution in Charophyta. *Indian Biol.* 2 : 11-19.

87. Sarma, Y.S.R.K and Ramjee, 1967. A new chromosome number in *Nitella tenuissima* (Desv.) Kuetz em RDW, *Caryologia,* 20 : 61-63.

88. Sarma, Y.S.R.K and Ramjee, 1969. Chromosome numbers in three taxa of Indian Charophyta, *J. Cytol. Genet.* 4 : 32-36.

89. Sarma, Y.S.R.K and Ramjee, 1971. Significance of chromosome numbers in Charophyta. A discussion. *Caryologia* 24 : 391-401.

90. Sarma, Y.S.R.K and S.B. Singh, 1974. Effect of gamma rays on the chromosomes of *Nitella flagelliformis, Cytologia,* 39 : 303-308.

91. Sarma, Y.S.R.K and S.B. Singh, 1977. Irradiation studies on the karyology of Charophyta-I. X-rays and gamma rays, *Cytologia* 42 : 279-290, 291-295.

92. Sarma, Y.S.R.Kand S.N. Tripathi, 1976. Effect of chemicals on some members of Indian Charophyta, I.*Caryologia* 29 : 247-262, 263-276.

93. Selvaraj, R. and D. Subramanian, 1980. Chemotaxonomical studies in some South Indian species of Rubiaceae. III, *All India. Bot. Conf.* Vol. 59, p. 41.

94. Sinha, J.P. and S. Ahmad, 1973. Chromosome numbers of *Nitella* Ag. From Bihar, *Proc. Indian Sci. Cong. Assoc.,* 319.

95. Sinha, J.P. and S. Ahmad, 1974. Karyological studies on *Chara vandalurensis* Sund. from India*Proc. Inter Symp. Taxo,* Madras, 38.

96. Sinha, J.P. and S. Ahmad, 1979. Cytological analysis in two species of *Nitella* Ag. from Bihar.*Proc. Indian Sci. Cong. Asso.* 26.

97. Sinha, J.P. and M.N. Noor, 1967. Chromosome numbers in some members of Charophyceae of Chhotanagpur Plateau, India. *Phykos,* 6 : 106-109.

98. Sinha, J.P and M.N. Noor, 1971. Studies on the karyology of some members of Charophyta of Bihar, Phykos, 10 : 112-117.

99. Sinha, J.P and B.N. Verma, 1970. Cytological analysis of the Charophytes of Bihar, *Phykos,* 9 : 92-99.

100. Sinha, J.P and B.N. Verma, 1969. A new chromosome count for *Chara braunii* Gm. *Proc. Indian Sci. Cong. Asso.* 275.

101. Subramanian, D. and I.V. Krishnamurthy, 1978. Chemotaxonomical studies in some species of Acanthaceae. *NationalAcademy Sci. Letters.* 142 : 53-55.

102. Subramanian, D., 1980. Chemotaxonomical studies in South Indian Polygonaceae. *J. Ind. Bot. Soc.* 50 : 119-122.

103. Subramanian, D.,1981. On new Charophytes from Tamilnadu, *Phykos* 20 (1&2) : 1-7.

104. Subramanian, D.,1981. Ecological studies of Charophyta. Abs. presented in *Seminar on Algae and their utilization at Madras,* p. 36.

105. Subramanian, D.,1982. Studies on Charophyta of Tamilnadu I. Taxonomic enumeration and chromosome counts of Charophyta from South Arcot District. *Phykos,* 22 : 18-22.

106. Subramanian, D., 1983. Two new charophytes from Tamilnadu, J. Indian Bot. Soc.62 : 211-214.

107. Subramanian, D.,and R. Ganesan, 1983. Chemotaxonomical studies of Charophyta from Tamilnadu. Abs. *All India Applied Phycological Congress at Kanpur,* p. 35.

108. Subramanian, D and R. Ganesan, 1983. Cytotaxonomical studies of Charophyta from Tamilnadu. Abs. *All India applied Phycological congress at Kanpur,* p. 37.

109. Subramanian, D, 1985. New varieties of *Nitella* from Tamilnadu. *J. Indian. Bot. Soc.* 64 : 306-311.

110. Subramanian, D, 1985. Evolution of quadriscutate antheridium in *NitellaJ. Indian. Bot. Soc.* 64 : 312-316.

111. Subramanian, D. and R. Ganesan, 1985. Cytotaxonomical studies of South Indian Charophyta. *Advances in Applied Phycology,*Kanpur, pp. 184-191.

112. Subramanian, D and R. Ganesan, 1985. Chemotaxonomical studies of South Indian Charophyta. *Advances in Applied Phycology,Kanpur,* pp. 107-112.

113. Subramanian, D., 1986. Four more new dioecious members of charophyta from Tamilnadu. *9th All India, Bot Conf. at Rajkot, Gujarat,* Vol. 65, Suppl. p. 1.

114. Subramanian, D., 1987. New varieties of *Nitella* from Western Glats of Tamil Nadu, Proc. *74th, Indian Sci. Cong. At Venkateswara Uni., Thiruppathi,* A.P. Part III, Abs. Sec. VI, Botany, p. 119.

115. Subramanian, D., 1987. Further chemotaxonomical studies of South Indian Charophyta I. *All India, Applied, Phycological Cong.,* New Delhi.

116. Subramanian, D., 1988. Biochemical taxonomy of South Indian Charophyta Abs. presented at *3rd International Phycological Cong. Monash Univ., Melbourne,* Australia.

117. Subramanian, D., 1998. Some interesting Charophytes from Tamilnadu. Late Abs. presented at *Seaweed Research & Utilis Association, National Symposium* held from 8 to 10, May at MaduraiKamarajUniversity, p. 8.

118. Subramanian, D., 1999. Some interesting members of Charophyta from Tamil Nadu. *Seaweed Research and Utilis,* 21 (1&2) : 105-111.

119. Subramanian, D., 1999. New Varieties of *Nitella* from Tamil Nadu. *National Symposium on recent trends in Marine Algal Research in India.* Dec. 27 to 29, Portonovo, Abs. No. 18.

120. Subramanian, D., 2000. Biosystematical studies in Charophytes. Spl. Lecture. *National Symposium in Phycology in the New Millennium* from 1 to 3, March. CAS in Botany, ChennaiUniv.

121. Subramanian, D., 2000. Charophytic flora of Natural ponds in Virudhachalam of Tamil Nadu. *National Symposium on Phycology in the New Millennium,* 1 to 3, March CAS in Botany, ChennaiUniversity.

122. Subramanian, D., 2002, *"Monograph on Indian Charophyta"* Bishen Singh Mahendra Singh Publications, Debradun.

123. Subramanian, D., 2016, *"Ecological adaptations of Indian charophyta"* Abs. International Conference on Advances in Algal Biotechnology VIT. University, Vellore, p. 19.

124. Subramanian, D., 2016, *"Charophytes* from Karnataka State, India. Abs. International Conference on Advances in Algal Biotechnology VIT. University, Vellore, p. 50.

125. Sundaralingam, V.S., 1959. "A systematic account of the South Indian Characeae" *Proc. Indian Academy Sci.* 12 : 7-51.

126. Wood, R.D. and Imahori, K., 1964. *A Revision of the Characeae* Parts I & II. Verlag van J. Cramer. Weinheim, Germany.

EXPLANATIONS OF PLATES AND FIGURES

Plate – 1: *Nitella furcata*. var. *mucronata. f. puduchattirensis.*

Plate – 2: *Nitella furcata*

Plate – 3: *Chara fibrosa*

Plate – 4: *Nitella hyalina - var. athanarensis*

Plate – 5: *Chara hornemannii*

Plate – 6: *Chara hornemannii f. norhoffiae*

Plate – 7: *Chara zeylanica*

Plate – 8: *Nitella terestris*

Plate – 9: *Nitella terestris*

Plate – 10: Figs. 1 & 2. *Nitella congesta var. filiformis*. 1 = A Plant ; 2 = A whorl enlarged. Figs. 3 to 8 : *Nitella hyalina*. 3&4 = Plants ; 5 = A whorl enlarged; 6 to 8 = 3 types of branchlets.

Plate – 11: *Nitella grandiformis* from Madras.

Fig. 1 to 3 = Plants ; 4 = A Plant slightly enlarged ; 5 = Axial node with branchlets; 6 = A branchlet enlarged; 7 = tip of branchlet.

Plate – 12: *Nitella lhotskeyi. f. hyalinoides.*

Fig. 1= Plant natural size; 2 = plant slightly enlarged; 3 = A portion enlarged; 4 = A whorl with a large branchlet and a small branchlet; 5 = dactyls.

Plate – 13: *Nitella lhotskeyi. f. minima.*

Fig. 1 = A plant, natural size; 2 = 2 whorls enlarged; 3 = large branchlet; 4 = small branchlet; 5 = A dactyl; 6 = Antheridium; 7 = Oogonium.

Plate – 14: *Nitella hyalina.* Thirunelveli Form.

Fig. 1 = A plant = 2 = A whorl enlarged; 3 = A whorl with a larger branchlet; 4 = medium sized branchlet; 5 = small branchlet.

Plate – 15: Figs. 1 to 5 = *Nitella hyalina.* Mavidanthal Form.

Fig. 1 = A Plant; 2 = Portion of plant enlarged ; 3 to 5 = 3 types of branchlets.

Figs. 6 and 7 = *Nitella dispersa.* Mavidanthal Form. Fig. 6 = Male Plant ; 7 = Female Plant.

Plate – 16: *Chara corallina.*

Figs. 1 to 4 = Form 1. Fig. 1 = A Plant; 2 = Sex organs; 3 = A branchlet ; 4 = branchlet tip. Fig. 5 = Form 2. A. Plant.

Plate - 17: Figs. 1 to 3 = *Chara canescens.*

Fig. 1 = A plant ; 2 = A branchlet; 3 = Two nodes, one with A wining cortical threads and another with branchlets.

Figs. 4 to 6 = *Chara fibrosa f. meenakshi.*

Fig. 4 = a plant ; 5 = Axis node; 6 = A branchlet.

Plate – 18: *Nitella leptostachys*. var. *viscosa* Puduchattiram type.

Fig. 1 = A plant enlarged; 2 ; sterile branchlet; 3 = Fertile whorl ; 4 = Dactyls = 5 = Corona cells; 6 = Oogonium; 7 = Antheridium ; 8 = Oospore membrane.

Plate – 19: *Nitella verticillata*. var. *kodaikanalensis*.

Fig. 1 = A plant ; 2 and 3 = whorls; 4 = Sex organs; 5 = Dactyl; 6 = Antheridium ; 7 = Oogonium.

Plate – 20: *Chara fibrosa. f. robusta*. Sankarapuram Form.

Fig. 1 = Plant ; 2 = A whorl with branchlets; 3 = branchlet of male plant ; 4 = branchlet of female plant ; 5 = An axial node with stipulodes; b = Antheridium; 7 = Oogonium.

Plate – 21: *Nitella grandiformis*. var. *hyalinioides*.

Fig. 1 = A plant; 2 = A vegetative whorl with axillary fertile whorl in the axis; 3 = fertile whorl. 4 = Dactyl.

Plate – 22: *Chara vulgaris*. var. *yercaudensis*.

Fig. 1 = A plant ; 2 = A portion of plant enlarged; 3 = A branchlet.

Plate – 23: *Nitella arechavaletae*. From Topslip.

Fig. 1 = A plant ; 2 = A sterile whorl with an axillary fertile whorl ; 3 = Fertile whorl enlarged; 4 = Antheridium ; 5 = Oogonium ; 6 = Dactyls.

Plate – 24: *Chara fibrosa*. var. *robusta*. Fig. 1 = A plant ; 2 = Main axis showing stipulodes, spine cells and branchlets; 3 = A branchlet with sex organs.

Plate – 25: *Nitella furcata*subsp. *mucronata.*

Fig. 1 = A plant ; 2 = A portion of the plant enlarged; 3 = dactyls; 4 = sex organs

Plate – 26: *Nitella translucens* var. *minima.*

Fig. 1 = A plant; 2 = Basal sterile whorl with a branchlet ; 3 = upper fertiler whorls; 4 = Dactyl;

Plate – 27: *N. translucens* var. *annamalaiensis.*

Fig. 1 = A plant ; 2 = portion of a plant with sterile whorls and fertile whorls enlarged ; 3 = Dactyls; 4 = Oogonium ; 5 = Antheridium.

Plate – 28: *Nitella flexilis.* var. *lancifolia.*

Fig. 1 = A plant ; 2 = A portion of plant with sterile and fertile whorls enlarged ; 3 = A dactyl.

Plate – 29: *N. flexilis.* var. *grandifolia.*

Fig. 1 = A plant; 2 = A portion of a plant with sterile and fertile whorls enlarged.

Plate – 30: *Nitella hyalina.* Form 1.

Fig. 1 = A plant; 2 = A whorl within mucilage clouds ; 3 to 5 = 3 types of branchlets; 6 = Antheridium; 7 = Oogonium.

Plate – 31: *N. hyalina.* Form – 2

Fig. 1 = A plant ; 2 & 3 = 2 whorls enlarged; 4 = Dactyl; 5 = Antheridium ; 6 = Oogonium.

Plate – 32: *N. hyalina.* Form – 3.

Fig. 1 = A plant; 2 = A whorl enlarged; 3 to 5 = 3 types of branchlets.

Plate – 33: *Nitella annamalaiensis.*

Fig. 1 = A plant enlarged ; 2 = A fertile whorl arising from the first node of sterile branchlet. 3 = sterile branchlet ; 4 = Fertile branchlet ; 5 = A dactyl.

Plate – 34: *Nitella verticillata.* From Sathamangalam.

Fig. 1 = A plant ; 2 = A plant enlarged ; 3 = Dactyls.

Plate – 35: *Nitella doidgeae f. congesta*

Fig. 1 = A plant; 2 = A whorl with a branchlet enlarged; 3 = dactyls; 4 = sex organs.

Plate – 35a: *N. tenuissima.*

Fig. 1 = A plant ; 2 = A whorl with a branchlet enlarged; 3 = Dactyls ; 4 = Antheridium; 5 = Oogonium ; 6 = Corona cells ; 7 = Oospore.

Plate – 36: *Nitella charoides*

Plate – 37: *Chara zeylanica* var. *robusta.*

Plate – 38: *C. socotrensis.* var. *incurvata*

Plate –39: *C. socotrensis.* var. *ebracteata*

Plate – 40: Photographs of

1. *N. charoides*, 2. *Chara zeylanica.* var. *robusta*;

3. *C. socotrensis.* var. *incuruata*;

4. *C. socotrensis.* var. *ebracteata.*

Plate – 41: *Chara zeylanica* var. *spinosa* from Sriperumpudur.

Plate – 42: 3 forms of *Chara corallina* and a form of *C. braunii.*

Plate – 43: *Tolypella dispersa.*

Fig. 1 = A plant enlarged with sex organs ; 2 = Dactyls; 3 = sex organs ; 4 = Antheridium ; 5 = Oogonium ; 6 = Oospore membrane wall ornamentation.

Plate – 44: *Chara vulgaris. f. glabra.*

Fig. 1 = Plant ; 2 = Axial node ; 3 = A branchlet ; 4 = Branchlet tip; 5 = sex organs.

Plate – 45: Figs. 1 to 4 = *C. vulgaris. f. simplex.*

Fig. 1 = A plant ; 2 = Axial node ; 3 = A branchlet ; 4 = sex organs.

Figs. 5 to 7 = *C. braunii. f. delicatula.*

Fig. 5 = Plant; 6 = A branchlet ; 7 = sex organs.

Plate – 46: *Chara vulgaris.* var. *kovertinensis.* Fig. 1 plant; 2 = Axial node; 3 = A branchlet; 4 = sex organs

Plate – 47: *Chara tomentosa.* var. gigantea.

Fig. 1 = A plant ; 2 = Axis with a whorl of branchlets; 3 = Axial node enlarged ; 4 = A branchlet ; 5 = sex organs.

Plate – 48: *Nitella umbellata.*

Fig. 1 = A plant ; 2 = part of plant enlarged ; 3 = A branchlet ; 4 = A dactyl.

Plate – 49: *Chara bauerii.* var. 1.

Fig. 1 = A plant ; 2 axial node with branchlets; 3 = A branchlet enlarged; 4 = Sex organs.

Plate – 50: *C. bauerii.* var. 2

Fig. 1 = A plant ; 2 = Axial node with branchlet whorl ; 3 = A branchlet enlarged ; 4 = sex organs.

Plate – 51: *Chara bauerii.* Var 3

Fig. 1 = A plant ; 2 = Axial node with branchlets; 3 = Axial node with stipulodes; 4 = A branchlet ; 5 = sex organs.

Plate – 52: *Chara leptopitys.* Var 3.

Fig. 1 = A plant ; 2 = Part of the plant enlarged; 3 = A branchlet with sex organs; 4 = sex organs.

Plate – 53: *Nitella bastinii*

Fig. 1 = A plant enlarged; 2 = A fertile whorl; 3 = A fertile branchlet with sex organs; 4 = Dactyls of fertile branchlet; 5 = dactyls of sterile branchlet

Plate – 54: *Nitella duthieae.*

Fig. 1 = A plant ; 2 = Tip whorls ; 3 = Branchlets ; 4 = Dactyls ; 5 = sex organs.

Plate – 55: *Nitella hyalina.* Form – 1

Fig. 1 = A plant ; 2 = A whorl enlarged ; 3 = 3 types of branchlets; 4 = A dactyl.

Plate – 56: *Nitella hyalina.* Form – 3.

Fig. 1 = A Plant ; 2 = A whorl enlarged; 3 = 3 types of whorls; 4 = sex organs.

Plate – 57: *Nitella hyalina*. Form – 4.

Fig. 1 = A plant ; 2 = A whorl enlarged ; 3 = A dactyl; 4 = sex organs.

Plate – 58: *Nitella congesta* var. *hyalinoides*

Fig. 1 = A plant ; 2 = Portion of plant enlarged; 3 = A whorl; 4 = A branchlet ; 5 = A dactyl; 6 = part of whorl with 100 gonium 7 = Oogonium.

Plate – 59: *Lychnothamnus barbatus. f. minima.*

Fig. 1 = A plant slightly enlarged; 2 = A plant more enlarged; 3 = sex organs.

Plate – 60: *Nitella lhotskeyi form* - 1

Figs. 1 and 2. *Nitella verticillata*. Fig. 1 = A plant; 2 = A plant enlarged. Figs. 3 to 5 = *Nitella lhotskeyi*; Fig. 3 = A plant; 4 = Portion of plant enlarged; 5 = A branchlet with sex organs 6 = sex organs.

Plate – 61: *Nitella lhotskeyi*. Form -2

Fig. 1 = A plant ; 2 = Portion of plant slightly enlarged ; 3 = Portion of plant enlarged; 4 = 2 types of branchlets ; 5 = sex organs.

Plate –62: *Chara globularis. From Patiala*

Fig. 1 = A plant ; 2 = A portion of the plant with main stem and a whorl of branchlets enlarged. 3 = A branch with 2 nodes enlarged.

Plate – 63: *Chara corallina from Ambala.*

A

INDEX

A

B

C

D

E

F

G

H

K

L

M

O

P

R

S

T

W

Y